Virtual Reality Audio

Virtual Reality Audio: A Guide to Immersive Sound in Virtual Experiences takes you inside the world of immersive sound in VR, revealing how it transforms music, film sound, and game audio through practical examples and rich aesthetic, technical, and historical insights.

Readers will discover the perceptual foundations of immersive sound and its development into an inextricable part of virtual experiences today, before diving into hands-on examples that explore cutting-edge applications of VR audio across multiple disciplines. This book is accessible for a wide range of readers, including audio professionals looking to expand into VR, visual and interactive media specialists seeking to harness the power of immersive sound, and multidisciplinary VR creators eager to build their skills in audio implementation.

With a blend of theory and applied practice, *Virtual Reality Audio* offers a comprehensive roadmap for anyone interested in the future of sound in virtual worlds.

Anıl Çamcı is a professor in the Department of Performing Arts Technology at the University of Michigan, where he leads the Worldmaking Lab. His work has received numerous awards internationally and has been featured in leading conferences and journals on immersive audio, extended reality, and human-computer interaction.

Jason Corey is a professor in the Department of Performing Arts Technology at the University of Michigan, where he teaches technical ear training and sound recording. He wrote the book *Audio Production and Critical Listening: Technical Ear Training* and is active as a recording engineer.

Audio Engineering Society

Audio Engineering Society Presents
www.aes.org

Modern Recording Techniques, 10th Edition
A Practical Guide to Modern Music Production
David Miles Huber and Emiliano Caballero

Recording Orchestra and Other Classical Music Ensembles, 2nd Edition
Richard King

Sound Reproduction, 4th Edition
The Acoustics and Psychoacoustics of Loudspeakers, Rooms and Headphones
Floyd E. Toole, Sean Olive and Todd Welti

Classical Recording, 2nd Edition
A Practical Guide in the Decca Tradition
Caroline Haigh, John Dunkerley and Mark Rogers

Handbook for Sound Engineers, 6th Edition
Edited by Glen Ballou and Doug Jones

Audio Effects: Theory, Implementation and Application
Joshua D. Reiss and Andrew McPherson

Virtual Reality Audio
A Guide to Immersive Sound in Virtual Experiences
Anıl Çamcı and Jason Corey

For more information about this series, please visit: www.routledge.com/Audio-Engineering-Society-Presents/book-series/AES

Virtual Reality Audio

A Guide to Immersive Sound in Virtual Experiences

Anıl Çamcı and Jason Corey

Routledge
Taylor & Francis Group
LONDON AND NEW YORK

Designed cover image: Anıl Çamcı

First published 2026
by Routledge
4 Park Square, Milton Park, Abingdon, Oxon OX14 4RN

and by Routledge
605 Third Avenue, New York, NY 10158

Routledge is an imprint of the Taylor & Francis Group, an informa business

British Library Cataloguing-in-Publication Data
A catalogue record for this book is available from the British Library

ISBN: 978-0-367-49899-3 (hbk)
ISBN: 978-0-367-49897-9 (pbk)
ISBN: 978-1-003-04802-2 (ebk)

DOI: 10.4324/9781003048022

Typeset in Sabon

by SPi Technologies India Pvt Ltd (Straive)

Access the Support Material: vraudio.org

Contents

Preface

Sound is essential to our experience of physical reality. It helps us communicate with each other, perceive and navigate the spaces we occupy, and stay alert to events happening around us. Auditory perception is a survival skill we regularly rely on and excel at effortlessly. We can detect minute variations in the pitch, dynamics, timbre, and localization of everyday sounds and react to them instantaneously. Sound is also a cultural phenomenon. We organize sounds into music to entertain ourselves, shift our moods, form social bonds, and cultivate cultural identities. Being such a vital component of the human experience, it is no wonder that sound is inextricable from simulations of that experience in virtual realities.

In the decade since the mid-2010s, VR technology has crossed a usability threshold and gained unprecedented consumer adoption. Today, virtual and, more broadly, extended reality technologies are rapidly emerging as a new frontier in media delivery and personal computing. This has cast a new light on spatial audio, which has a mutually beneficial relationship with VR. On the one hand, spatial audio significantly contributes to the sense of immersion and presence in VR. The inherently volumetric nature of sound and our acute ability to perceive it in space can make or break VR experiences. On the other hand, VR systems, and particularly the spatial tracking technology built into them, improve the performance of spatial audio techniques. Furthermore, VR puts spatial audio in a multisensory and interactive context, where our auditory perception is reinforced by stimuli in other modalities.

While sound design for VR is informed by existing practices in media production, it requires new tools and techniques, as well as a fundamental rethinking of the relationship between sound, space, and the user. In other words, VR audio practitioners not only need to learn new skills but also adopt new aesthetic frameworks to take full advantage of this novel creative medium. While this may come as a daunting task, the emerging nature of VR implies ample opportunity for impact and innovation, even for those new to the field.

In this book, we take a comprehensive look at VR audio, weaving a thread across the physics of sound in space, the human perception of auditory phenomena, the computational implementation of spatial sound, and the application of immersive audio to VR applications. By doing so, we aim to provide our readers with a lasting understanding of audio in the rapidly-evolving landscape of VR technology. To that end, we combine audio theory and techniques with practical applications in music, film, and game production for VR, offering a wide-ranging guide to immersive audio supported by workflow descriptions, online demos, and sample media. We frame this with both historical perspectives and aesthetic analyses of immersive audio.

This book is intended for a broad range of readers: those with a background in audio looking to expand their practice into VR; professionals in visual and interactive media seeking to understand and leverage the critical role of sound in immersive applications; and individual creators who work across all aspects of VR design and want to gain experience in audio implementation. While prior audio knowledge may be helpful to navigate some sections, we don't assume an expertise in either audio or VR—only a deep curiosity about the subject.

Throughout the book, we use accessible language to connect key concepts across acoustics, auditory perception, spatial audio, and VR technology. We support this with historical and aesthetic perspectives on immersive media as we chart the evolution of VR from early ideas to contemporary applications. Furthermore, we present case studies that situate VR audio in an interdisciplinary framework and draw from practices in film, games, and interaction design. We demonstrate varied tools and techniques, focusing on workflows that are unique to immersive media and those that have not been extensively covered in prior literature. In these case studies, we utilize open-source and free-license software wherever possible to allow our readers to download and follow along with our example projects. Our aim is to assist a diverse range of readers—from students and educators to enthusiasts and professionals—in expanding their grasp of the immersive medium and contributing to its growth in new and exciting directions.

In Chapter 1, we explore the history of VR with an emphasis on developments in the past decade. We then shift our focus to audio and launch our book-long conversation on the critical role of immersive audio in VR experiences and how we can best utilize it.

In Chapter 2, we survey the history of immersive audio from early experiments in the nineteenth century all the way up to modern surround sound and soundfield techniques. We explore the development of audio technology for cinema, which has traditionally spearheaded consumer applications of spatial audio.

In Chapter 3, we discuss fundamental aspects of sound propagation, acoustics, and auditory perception to lay the groundwork for simulation of spatial sound in VR. We investigate the perceptual cues that inform our localization of sound in space and how these contribute to our sense of place and immersion.

In Chapter 4, we focus on techniques that allow us to simulate propagation and localization of sound computationally. We discuss how sounds can be captured or rendered for immersive delivery, looking at spatial microphone and encoding techniques. We also review different methods of spatial audio diffusion through loudspeaker arrays, as well as headphones.

In Chapter 5, we explore the role of audio in VR film by drawing from traditional practices in film production and expanding into the idiosyncrasies of designing sound for fixed immersive media. We highlight the relationship between the existing roles, tools, and workflows in film production and those that are specific to VR. We present two case studies that deal with different aspects of VR film, demonstrating some of the technical and aesthetic considerations presented earlier on.

In Chapter 6, we look at another prominent domain of immersive media, namely VR games. We discuss the fundamental concepts of VR audio in interactive experiences and showcase modern tools used in the application of these concepts to game and interaction design. We present an additional set of case studies that explore the role of spatial and procedural audio in two of the leading game engines today.

Finally, in Chapter 7, we take a broad look at the present and future of VR audio to chart a path for creative practices moving forward. We touch upon a wide range of topics from the future of entertainment media to ethical considerations in VR. We summarize the key takeaways from the book and relate them to developments in the industry and numerous disciplines that have a stake in the evolution of VR.

Bridging the gap between theory and practice, we aim for this book to serve as a compelling and informative resource for both novice and experienced practitioners who seek to harness the unique potential of VR for audio creativity and to use sound to shape VR experiences. As you embark on your journey through this book, we encourage you to experiment boldly, push the boundaries of your creativity, and engage with the ever-expanding possibilities of virtual reality audio.

Acknowledgements

We are thankful for the amazing community of artists and researchers in extended reality (XR), who have not only motivated this book but also informed our work on it. We would also like to thank our students, who have given us unique insights into how XR is learned and put into creative practice. Working with them over the years has helped us shape the scope and structure of this book.

We would like to thank Evie Evans, Emily Tagg, Hannah Barnes-Rowe, and Shannon Neill at Routledge for all their help and guidance throughout our writing process. We would also like to thank our external reviewers, whose feedback provided us with new perspectives on the book and helped us make it a better contribution to the field.

We are grateful to Len Moskowitz of Core Sound LLC and Gary Elko of MH Acoustics for granting us permission to use photos of Ambisonic microphones from their respective companies. We also thank Philip Popp for his insights on MetaSounds in Unreal Engine.

We owe a debt of gratitude to our collaborators who contributed to the case studies presented in this book. In the first case study in Chapter 5, we had the pleasure of working with our friend and colleague Andy Kirshner, his production crew, including Juliet Hinely, Laura Pazuchowski, Jean Nardone, Linda Long, and Brenna K. Murphy, and the production staff at the Duderstadt Visualization Studio, including Jacques Mersereau, Catherine Miller, Patterson McKinney, and Matthew Girard. We are also deeply grateful to the four actors, Emily Sutton-Smith, Malcolm Tulip, Caroline Helton, and John Lepard, who delivered brilliant performances despite the unconventional technical setup we subjected them to.

In our second case study, we are grateful for the outstanding work of the University of Michigan's Percussion Ensemble, comprising Xin Yi Chong, James Koo, Brenna Lantz-Lewis, Rachel Richards, Jonathan Wentzel, and Grant Johnson, under the leadership of percussion faculty Ian Antonio and Doug Perkins. We also thank composer Carolyn Chen for the wonderful piece of music that the ensemble performed for the case study. We are grateful to our colleague Aaron Kippola for his insight and support during this recording session. Additionally, we would like to thank our students Eric Cerruti Lisboa de Oliveira, Eva Choi, and Eli Faber for their assistance with the case studies.

Anıl Çamcı: I would like to thank my co-author, Jason Corey, who has been a thoughtful and dedicated creative partner. It has been a pleasure to collaborate with him on this project. I am indebted to my family for the unwavering support that made my work on this book possible. My father, Akın Çamcı, passed away during the writing of the book, and I would like to dedicate my work to his memory. My mother, Meral Çamcı, has tirelessly championed me and this project even though she doesn't consider herself among the book's core audience. My brother, Alp Çamcı, who is a tech aficionado, tried VR for the first time as we began writing, and his enthusiasm as a novice user was an early dose of encouragement. Finally, I want to thank my wife, Zeynep Özcan, for always standing by me and my work without hesitation. I could not have completed this project without the love, support, and inspiration she gives me every day.

Jason Corey: I would like to thank my collaborator and co-author, Anıl Çamcı, for many fruitful discussions about VR audio and for his leadership on this book project. He provided countless insightful and invaluable comments and suggestions during the writing and editing process. My work on this book would not have been possible without the love and support of my family. Thank you to my family for encouraging and supporting my professional aspirations and for allowing me the time and space to finish this project.

Chapter 1

Introduction

A New Frontier for Audio Creativity

In this chapter, we provide an overview of recent developments and emerging trends in virtual reality (VR) to set the stage for our discussion on VR audio. We begin by examining both historical and modern VR technologies to contextualize the revitalization of the VR industry throughout the 2010s. We follow this with our projections for its future evolution, particularly the convergence of extended reality (XR) technologies. We then discuss the influence of sound on presence, immersion, and agency in VR—a theme that we will expand upon throughout the book. We touch upon advances in VR audio in terms of tools, techniques, and applications with a particular focus on creativity support for spatial audio. Finally, we identify emerging challenges and opportunities for content creators and highlight the pivotal role of VR audio in shaping cutting-edge immersive experiences.

1.1 A Condensed History of VR

Creating representations or abstractions of reality has been a long-standing obsession of humankind, dating back to the cave painting of a Sulawesi warty pig more than forty-five thousand years ago (Brumm et al. 2021). Many millennia later, Plato conceived the Allegory of the Cave, which was not only a statement on the role of scholarship in forming one's worldview but also an early thought experiment on what one perceives as reality. In another famous rumination about the thin line (or lack thereof) between reality and virtuality, Descartes pondered how dreams are sometimes indistinguishable from waking life, suggesting that life itself could be a dream. These are just a few examples of humanity's desire to interpret, question, and challenge our understanding of reality. Historically, the arts have been one of the foremost avenues for satisfying this desire. Many art forms are intrinsically engaged in creating portrayals or modifications of reality. Paintings can depict shapes, objects, and places that are inspired by, yet nothing like, the shapes, objects, and places that we have seen before. A novel can carry us into a world separate from the one we inhabit. Music can evoke both memories of past experiences and visions of those yet to be lived. Through these diverse means, the arts already engross us in virtual realities.

1.1.1 Early Implementations

When we seek a more explicit interpretation of VR, we find numerous technical milestones throughout the last century that have led to the development of VR technology as we know it today. In an example from the early twentieth century, the View-Master

DOI: 10.4324/9781003048022-1

utilized the principle of stereoscopy to display per-eye images of a scene from slightly shifted perspectives to evoke a sense of visual depth. This same principle was applied in early 3D cinema, where viewers wore anaglyph glasses that cast two different images of the same scene in each eye, similar to how modern 3D cinema works. In 1960, the cinematographer Morton Heilig patented a device called the Stereoscopic Television Apparatus for Individual Use (Heilig 1960), which is widely recognized as an unrealized precursor to modern head-mounted displays. A few years later, Heilig developed the Sensorama, an arcade machine that combined stereoscopic film with stereo sound, smells, wind, and vibration. In 1968, the computer scientist Ivan Sutherland introduced the first head-mounted display with head tracking and 3D computer graphics. The following year, the artist Myron Krueger began to develop a series of works that placed users in computer-generated interactive virtual spaces using computer vision. Krueger later coined the term "artificial reality" to characterize these works (Krueger 1977).

In 1977, Dan Sandin and Tom DeFanti created the first data glove at the University of Illinois Chicago's Electronic Visualization Lab (EVL). An evolution of this technology, aptly named DataGlove, was developed at VPL Research founded in 1984 by Jaron Lanier, who coined the term "virtual reality" (Lanier 1992). VPL Research also developed a full-body version of the DataGlove as well as a head-mounted display called the EyePhone. In 1992, Carolina Cruz-Neira, Dan Sandin, and Tom DeFanti of the EVL introduced the first Cave Automatic Virtual Environment (CAVE), which provides a room-scale stereoscopic VR experience through surround video projection and motion tracking (Cruz-Neira, Sandin, and DeFanti 1993).

Beyond these advances at research institutions, there were also commercial efforts to popularize VR throughout the 1990s, with popular video game console manufacturers attempting to build the first consumer VR headset. In 1991, Sega began developing Sega VR, which resembled today's VR systems but never moved beyond the prototype phase. In 1995, Nintendo's Virtual Boy became the first video game console with stereoscopic graphics but failed to achieve commercial success (Boyer 2009).

1.1.2 Modern VR Systems

As we can tell from these examples, advances in computing throughout the twentieth century have led to repeated attempts at realizing VR every decade or so with varying degrees of success. However, it wasn't until the 2010s that commercially viable products began to emerge. In 2010, Microsoft introduced the Kinect depth camera, which extracted a point-cloud image of its surroundings and performed skeletal tracking to detect users within its field of view. Although the Kinect was initially conceived as an input device for the Xbox video game console, it has prompted numerous applications across a range of disciplines from music to medicine, allowing bodily movements to be mapped into 3D virtual environments. Based on the Kinect, Microsoft developed a proof-of-concept system called the IllumiRoom, which used projection to augment the user's play area in the style of a CAVE system (Jones et al. 2013). A device much smaller than the Kinect, called the Leap Motion, utilized a similar depth-tracking technique to detect the user's hands as an input mechanism.

Then in 2012, we witnessed the emergence of the latest and most promising wave of VR technology. Joined by game industry veteran John Carmack, entrepreneur Palmer Luckey introduced the first prototype of the Oculus Rift, a head-mounted VR display.

Leveraging recent advances in low-persistence display technology, the Rift offered relatively high resolution and refresh rate, overcoming some of the technical bottlenecks that impeded the growth of past VR technologies. Furthermore, the developer kits for the Rift were priced at a reasonably affordable point, which created excitement among the developer community (Gear 2013).

Around the time Oculus released the second developer kit for the Rift, Google introduced the Cardboard system, which allowed the integration of touchscreen phones into a cardboard headset. This made it possible for phones to be used as VR displays with three degrees of freedom (3DoF). While 3DoF systems capture rotational movement (i.e., changes in the pitch, roll, and yaw of the head in this case), they do not track positional information (i.e., where the head is in 3D space). As a result, while the user can rotate their head to look around in this type of VR, they can't move their body to navigate virtual space. Six degrees of freedom (6DoF) systems, on the other hand, track both rotational and positional changes, allowing room-scale movement in physical space to be mirrored in virtual space.

The first consumer version of the Rift was introduced in 2016 with an external camera system for 6DoF tracking and a pair of Touch controllers for hand-based interactions. Both the developer and consumer versions of the Rift were tethered systems, wherein the graphics processing is carried out on a separate computer before the result is delivered to the headset. Furthermore, both systems relied on outside-in tracking, where external cameras track the rotation and position of the headset so that an appropriate view of the virtual space can be displayed to the user. The same year, HTC introduced the Vive VR system in cooperation with Valve Corporation. Like the Rift, the Vive relied on outside-in tracking with external base stations and included hand-held controllers. Also in 2016, Sony introduced the PlayStation VR system, which has sold over two million units within a year, marking an unprecedented expansion of the VR user base.

In 2017, the Rift was followed by the Oculus Go, a more affordable headset that offered 3DoF tracking and on-device processing. While this allowed untethered operation, it was also limited in terms of processing power due to the mobile chipset embedded in the device. A year later, HTC introduced the Vive Pro, which improved on the original Vive with higher display resolution and additional features like a secondary camera on the headset to extract a depth image of the outside world.

The year 2019 saw the launch of numerous new VR systems. Oculus introduced the Quest line of headsets with a more capable mobile chipset and 6DoF inside-out tracking. In inside-out tracking, the system relies on an array of cameras on the headset to detect planar surfaces to track its surroundings. This allows the user to experience VR in room scale without being limited to a play area tracked by external sensors. Alongside the Quest, Oculus also released the Rift S, which relied on the same inside-out tracking principle but with the tethered operation of the original Rift. The same year, Valve released their own headset, called the Index, alongside their highly acclaimed VR game *Half-Life: Alyx*. The Index shared several features with the Vive but introduced additional degrees-of-freedom in its hand-held controllers and improved on the display quality. A few months later, HTC launched the Vive Cosmos series with inside-out tracking and standalone operation, positioning the headset as a direct competitor to the Quest. Before the end of 2019, Oculus introduced the Link capability for the Quest with a firmware update that allowed the headset to be tethered to a computer much like the Rift S. As a result of this overlap in functionality, the Rift series was discontinued with the introduction of the

Quest 2 in 2020. Improving on the screen resolution of the original Quest, the Oculus Quest 2 became one of the best-selling VR headsets according to data gathered from VR application marketplaces (Feltham 2021).

Soon after, the Oculus branding was discontinued by its parent company, Meta, starting with the 2022 introduction of the Meta Quest Pro headset, which featured eye and face tracking and introduced new Touch controllers with built-in cameras for 6DoF tracking. The Quest Pro was practically superseded in 2023 by the Meta Quest 3, which offered a higher-quality display and improved processing units at a more affordable price point. As of this writing, Meta Quest 3 is the flagship headset of the company, with the Quest 3s being their entry-level offering, first introduced in late 2024.

1.1.3 Convergence of Extended Realities

Parallel to developments in VR, the twenty-first century has seen significant advances in augmented reality (AR) and mixed reality (MR). As VR, AR, and MR technologies rapidly evolve, the distinctions between these paradigms are becoming increasingly blurred. AR and MR are particularly challenging to distinguish due to differing interpretations among practitioners (Milgram and Kishino 1994; Speicher, Hall, and Nebeling 2019). While some define AR as a subset of MR, others use the two terms interchangeably. Another common view differentiates AR as the overlay of virtual elements on the physical world and MR as the integration of virtual and physical objects that can interact with each other. Despite these conceptual and technical nuances that demarcate these paradigms, they both rely on detecting planar surfaces (or visual markers) to augment the physical world with virtual objects that persist spatially as the user's perspective shifts.

In 2016, Microsoft introduced the HoloLens mixed-reality headset, which employed an array of depth cameras and sensors to detect its surroundings and track the user's hands to establish a shared space between the user, their physical environment, and virtual objects. Two years later, Magic Leap introduced the Magic Leap One mixed-reality system. Unlike the HoloLens, which housed a computer inside the headset, the Magic Leap One included a tethered external processing unit. The latter also offered a combination of hand- and controlled-based user interactions. In 2019, Microsoft released the HoloLens 2, which improved on the original headset with a wider field of view, higher display resolution, and increased brightness.

Although there is ongoing debate on how to categorize immersive experiences between VR, AR, and MR, we have also been witnessing a convergence of these fields as they rely increasingly on common tracking, display, and interaction technologies. For instance, the plane detection employed in VR headsets with inside-out tracking was originally developed for markerless AR applications. Although the cameras built into these headsets for plane detection have been relatively low-resolution, it is becoming increasingly common for new VR systems to incorporate improved cameras to support MR applications, wherein the outside world is more prominently integrated into virtual space. For instance, in 2024, Apple introduced the Vision Pro, which is equipped with an array of cameras that can display a relatively high-resolution view of the outside world inside the headset. Although the Vision Pro fully occludes the user's view, Apple refrains from labeling it as a VR headset due to the central role of its see-through features in differentiating this product from other VR systems. Similarly, Meta's Quest Pro, 3, and 3s headsets, as well as their Horizon operating system for these devices, growingly integrate the outside world in virtual experiences.

The primary mode of interaction in most MR systems is hand tracking, which employs cameras built into the headset to detect the user's hand as an input mechanism. In recent years, this technique has been integrated into VR systems as an alternative to controller-based input. It is reasonable to expect the overlap between VR, AR, and MR technologies to expand towards systems that blur the boundaries between virtual, augmented, and mixed realities. This is why the industry has adopted the use of Extended Reality (XR) as an umbrella term that covers all these different kinds of immersive media. While it's difficult to predict which terminology will prevail in the coming years, there is a clear trend towards technologies that unify these seemingly distinct paradigms.

1.2 The Role of Audio in Immersive Experiences

Amid these rapid advances in VR technology, we are already confronted with fundamental questions about VR that will remain relevant, no matter which features emerge next: What are the intrinsic qualities of a VR experience? What makes this experience immersive? And how do we interact with it? Previously, we suggested the following as the pillars of a VR experience: 1) presence of virtual elements in either fully virtual or augmented reality settings; 2) persistence of a shared space between the user and these virtual elements; 3) the ability of the user to interact with this space; and 4) the stimulation of as many senses as possible (Turchet, Hamilton, and Çamcı 2021). While the first pillar offers a relatively straightforward characterization of VR today, it acknowledges the convergence of immersive media paradigms, as we discussed in the previous section. The second pillar emphasizes the anchoring of the user's embodied experience in virtual space through spatial tracking systems. The element of interactivity in the third pillar can range from a dynamic view of the virtual space via head tracking to room-scale experiences involving interactive virtual objects and agents. The fourth pillar has been central to many early conceptions of VR, dating back to Heilig's Sensorama, which addressed senses of smell, touch, and movement, alongside sight and hearing. In 1965, Ivan Sutherland laid out a vision for a "mathematical wonderland," a virtual environment rendered entirely by a computer (Sutherland 1965). He argued that the ultimate display would have to serve as many senses as possible despite the hurdles to achieving it back then: "So far as I know, no one seriously proposes computer displays of smell, or taste. Excellent audio displays exist, but unfortunately we have little ability to have the computer produce meaningful sounds." Since 1965, there have been significant advances in multimodal interfaces and digital computing. Today's computers can not only produce meaningful sounds but also spatialize them with high precision in real time.

In Chapter 3, we will explore how we perceive and localize sound in space. This will help us formulate a foundation for how certain tools and techniques came about in spatial audio. But even at a rudimentary level, it is easy to conceive why our perception of sound is an inherently immersive experience. Unlike our visual system, which has a limited field of view, our auditory system can pick up sounds from every direction, albeit with varying degrees of precision. Unless we cup or plug our ears, we do not have a biological mechanism to block our hearing the way we can use our eyelids to block our vision. We are constantly aware of and immersed in the auditory space around us. In any given environment, auditory cues give us a sense of place, scale, and orientation. We rely on these cues to gather information from our surroundings and navigate our daily lives. An increasingly loud engine noise tells us a car is approaching; the ding of a doorbell tells us someone is at the door; the constant drone of crickets or the rustling of trees in the wind are common markers of being outdoors; the reverberation in a building gives us a sense of its size even before we can fully see its dimensions. Beyond verbal cues with an explicit communicative function, every sound we hear helps us engage with our surroundings in various ways. This is also why audio plays a fundamental role in establishing a sense of presence in virtual environments; our sensory experience of the real world serves as a benchmark for interpreting these environments.

Sound has a well-established and extensively studied role in visual media such as films and video games, which we will focus on in Chapters 5 and 6, respectively. In addition to character dialogue, visual media employs a range of diegetic and non-diegetic sounds. Diegetic sounds are those that belong to the implied universe of a scene. For instance, the sound of a character's footsteps as they walk through a corridor is diegetic because it occurs within the film's world and is audible to the characters inhabiting it. These sounds can significantly affect how convincing and impactful a scene can be. On the other hand, the musical score of a film or a narrator's voice are non-diegetic because these sounds exist in a layer external to the story world. Non-diegetic sounds can serve various functions from setting the mood with music to conveying the storyline through narration. The first-person and dynamic perspective that the user gains in VR creates opportunities for blurring the traditional distinctions between diegetic and non-diegetic elements. The user can be situated in any role from a passive spectator to an active character. Moreover, this role can shift over time. For example, a musical track might initially play without any spatial cues, giving the impression of being external to the story world in the style of a non-diegetic score. This track could then be turned into a spatialized source (e.g., attached to a radio in the scene) that is now heard relative to the user's position and orientation, creating a link between the extra-diegetic layer of narration and the diegetic world of the virtual experience. These types of transformations can be used for narrative, interactive, or navigational purposes in VR.

Although existing theories and techniques of sound for visual media fundamentally inform our practices in VR audio, producing sound for VR introduces novel considerations and applications tied to the inherently immersive nature of auditory phenomena. Most obviously, sound can instill a sense of space in VR, just as it does in the real world. It can provide cues that indicate how vast or small a space is, how that space is designed and populated, and how we are situated in it. Such cues are essential for creating a sense of realism in VR. Even a sound as subtle as room tone (i.e., the sound of a quiet room) can be effective in making the user feel present in a virtual environment. This, for instance, is why the home screen in Meta VR has a faint noise that mimics an open-air soundscape rather than being completely silent. From there, more complex room and spatial effects, such as reverberation, absorption, occlusion, and distance modeling, can enhance the sense of immersion and plausibility in virtual spaces. Furthermore, these effects can contribute to the overall mood of an environment. By manipulating the virtual acoustics of a room, it can be made to feel more open, bare, cluttered, or even claustrophobic. Sound can also be used to give the user a sense of movement in VR. Especially in experiences that involve virtual navigation, moving sound sources can help evoke or amplify a relative perception of movement through space.

Beyond establishing presence, mood, and senses of space and movement, sound also serves more foregrounded functions in VR. For instance, audio cues can direct the user's attention to events or objects outside their field of view. Since the user has the agency to look wherever they choose in VR, they may miss objects or events that are vital to their experience. This can be particularly important in time-based and narrative experiences, where the progression of the story may be contingent on the user's awareness of these elements. In such cases, the designer can use sounds to alert the user to objects and events, either through diegetic cues (e.g., localized sounds emitted from the location of an object of interest) or non-diegetic cues (e.g., verbal instructions that direct attention to an object of interest). Throughout the book, we will continue to explore the substantial impact of sound on user experience in VR.

1.3 Creativity Support Tools for VR

In Chapters 5 and 6, we will focus on two primary types of VR content: fixed content based on recorded media, and interactive content generated computationally. VR film, for instance, is an example of fixed content, where video captured with a 3D camera is delivered through a VR headset, presenting a panoramic view of the real world in an immersive context. In contrast, computationally generated VR content often involves computer graphics. If traditional film can be seen as a precursor to VR film, then video games and animated films are precursors to computationally generated VR content. Although traditional video games are presented on 2D displays like computer monitors, many games leverage three-dimensional virtual environments that the user can explore through first- or third-person characters. Therefore, the tools and techniques that have been developed for game design naturally find their way into VR design. Today, most VR experiences are implemented with game engines such as Unity and Unreal. The modern game engine is an integrated development environment that combines 2D and 3D graphics, audio, physics simulation, animation, scripting, and debugging, among other functionality necessary to implement video games.

Most game engines offer built-in audio functionality to load and mix audio files, process these with audio effects, and spatialize them in 2D or 3D space. The Unreal game engine has recently introduced an audio graph system called MetaSounds, which can perform sample-accurate synthesis and processing of audio in real time, significantly broadening the audio capabilities of the engine. Such functionality can also be further expanded with external packages and custom libraries.

Besides using the built-in audio features of a game engine, game audio designers often employ audio middleware, such as Wwise and FMOD, to author interactive audio for games. These tools are used for handling, processing, and scheduling audio events in games. Although these are external tools that run outside the game engine, the audio events in middleware are triggered from within the game. These tools feature user interfaces and workflows similar to digital audio workstations (DAWs) and typically offer more advanced sound mixing capabilities than game engines.

There is also an extensive selection of domain-specific tools for sound spatialization. As we will discuss in Chapter 2, spatial audio has a rich history, dating back to the early days of sound reproduction in the late nineteenth century. While experiments in spatial audio in the days of analog electronics have led to hardware controllers like Pierre Schaeffer's Pupitre d'Espace, advances in digital computing brought about software systems that can render and move sounds in space. One of the most comprehensive tools in this field is IRCAM's sound spatialization suite called Spat (Carpentier, Noisternig, and Warusfel 2015), originally designed for Max/MSP but now available on other platforms and as a standalone application. Software tools like Spat implement one or more spatial audio techniques that allow the user to take recorded or live sound sources and spatialize them through a range of output configurations, including headphones and loudspeaker arrays. In Chapter 4, we will discuss some of these tools and techniques in greater detail.

Inviso

In this book, we will employ an online immersive audio design tool called Inviso (Çamcı et al. 2017) to demonstrate various concepts and techniques. The desktop version of Inviso, which can be accessed at inviso.cc, runs in a web browser, without requiring software installation. You can test the examples presented in the book and even design your own virtual sonic environments using Inviso. Upon launching the application in your browser, you will see an overhead view of an empty scene. You can press the add button—shown as a plus sign in the top right—and click anywhere to create spheres that emit omnidirectional sounds. You can also populate these spheres with cones that emit unidirectional sounds. You can press the add button and click-and-drag your mouse to draw zones with nondirectional sounds. You can move the listener node—represented with a blue head model—in the scene with W, A, S, D keys or with your mouse. You can add 3D motion trajectories to the spheres and the listener node. Aerial View allows you to move objects and trajectory control points horizontally. Tilt the scene by click-and-dragging anywhere outside the sound objects to enter Altitude View, where you can move objects and trajectory control points vertically. The Zoom In mode on a sphere allows you to fine tune the cones attached to it. You can also control each element parametrically using the graphical user interface (GUI) window that appears when an object is selected. You can pause individual elements or the entire scene. You can duplicate objects and zones. You can also undo or redo your actions. You can import any audio file from your system or use the built-in search box to find sounds on freesound.org. When you are done, you can export your scene for later use or sharing.

1.4 Challenges and Opportunities for Creatives

The future of a rapidly evolving medium like VR can be hard to predict. Frequent iterations in technology continually reshape how we engage with virtual environments. For example, modern VR systems have largely transitioned from outside-in tracking to inside-out tracking within just a few years. As a result, "play areas" have become much less rigid than those that can be defined with outside-in tracking. Moreover, many VR systems now integrate hand tracking for natural interactions that reduce the learning curve associated with hand-held controllers.

Another emerging trend in VR technology is the incorporation of eye tracking within the headset. Eye tracking can not only improve visual features like focus and depth of field but also enable new forms of user interaction based on gaze. Sony's PlayStation VR 2 system has adopted this feature for foveated rendering, where areas of a game that the user is directly looking at are rendered at a higher resolution than those in their peripheral vision. This allows the console to maintain high frame rates while providing the user with greater visual quality. Apple's Vision Pro employs gaze tracking for both foveated rendering and for interacting with the virtual environment.

In recent years, numerous features that improve the user experience in VR have been introduced in rapid succession. With each new VR system, we observe advances in frame rate, display resolution, and field of view, all of which significantly impact usability in VR. VR systems are also becoming lighter and more comfortable, making VR experiences less burdensome and more casual. Despite all this progress, we can still characterize modern VR systems as early manifestations of a new medium, akin to CRT televisions and dial-up modems. They offer us a glimpse into the exciting future of a new technological paradigm but suffer from many shortcomings. Today's VR headsets are bulky and can get warm easily, making them uncomfortable to use for extended periods of time. They either need to be tethered to a computer with powerful graphics processors or compromise on visual quality when used in stand-alone mode. Issues like the screen-door effect, where a user can see the borders between pixels, have yet to be fully resolved. Some of the modern headsets offer a limited field of view which can interfere with immersion. And the spatial tracking systems built into VR headsets and controllers can be prone to delays and disruptions.

Advances in audio hardware in VR systems have been relatively meager. The Oculus developer kits that marked the beginning of the latest wave of VR systems did not have any audio functionality. The first commercial version of the Rift came with built-in headphones, which, despite not being of high quality, allowed developers to integrate sound into their designs more seamlessly. With the introduction of the Quest series, these headphones were replaced with built-in speakers. While these can be even more limited in terms of audio fidelity, the headsets often include an audio output jack that allows the user to connect their own headphones. Looking at the past decade, it is not hard to discern that the VR industry has focused most of its hardware development efforts on display technology and graphics. Seeing that visual fidelity had traditionally been a major impediment to usability in past VR systems, this bias is reasonable. However, audio implementations in VR systems could certainly benefit from hardware features such as dedicated audio processing chips and better options for built-in audio input and output.

Regardless of how today's VR technology can be improved, it has already reached a level of growth and user adoption that none of the previous iterations of this technology had achieved. Today, millions of users own VR equipment across various platforms. Leading technology companies have positioned VR and, by extension, other extended reality technologies as a new frontier in entertainment and social engagement. However, modern VR is far from being considered a conventional media platform. One of the main bottlenecks in the mass adoption of VR is the limited availability of content that offers compelling immersive experiences to a broad range of audiences. Surveys show that the lack of content has been a primary barrier to the mainstream adoption of VR ("Augmented and Virtual Reality Survey Report" 2020).

While an inherent solution to this issue is for more people to create VR content, there are distinct creative and technical hurdles to this. Even though building VR experiences shares some similarities with traditional audio, video, and game production, it often demands new skills and perspectives specific to the immersive medium. In Chapter 5, we will identify areas where VR production strays from traditional practices in music and film. We will discuss some of the conceptual implications of producing content for a medium where the user is fully surrounded by the material and, in many cases, has a greater degree of agency over their experience. Additionally, the current VR workflows are not purpose-built for the immersive medium. We use film editing tools like Adobe Premiere and game engines like Unity to develop VR experiences. In Chapters 5 and 6, we will focus particularly on the audio design workflows in these tools. For instance, in game engines, audio implementations have traditionally been an afterthought with audio features delivered through user interface paradigms conceived for visual design. Therefore, these interfaces are rarely tailored for audio design. On the other hand, the software tools that are domain-specific to spatial audio design often come with a relatively steep learning curve. While the level of detail in these tools gives expert users granular control over sound spatialization, it can also prohibit novice users from engaging with spatial audio.

Another challenge in VR content creation is the need for software programming skills. Game engines often require proficiency in languages like C++ and C# to efficiently implement interaction mechanics and dynamic behavior. Additionally, VR companies offer native application programming interfaces (APIs) for building experiences specific to their platforms. These APIs can require an even greater degree of programming expertise and the content created with these tools may be locked to the platform they are designed for. This points to another challenge in modern VR production, namely platform fragmentation. Due to a lack of industry standards, creating content for two separate VR platforms can look vastly different. As a result, VR experiences are often not cross-platform and require substantial reworking to run on different systems. Open-source projects such as OpenXR and VRTK (Virtual Reality Toolkit) play a substantial role in addressing platform fragmentation in VR. However, these solutions also require advanced programming skills and can be susceptible to incompatibilities stemming from hardware or firmware updates from independent manufacturers.

An aspect of modern VR design that can be viewed as both a challenge and an opportunity is the constantly evolving nature of best practices in this field. As our collective understanding of VR as a media platform matures, we discover new methods to create more comfortable, engaging, and immersive experiences. For instance, how a user navigates a virtual scene and how their locomotion is reflected in VR can be a cause of motion sickness that prevents them from spending extended periods of time in VR. Every year, the industry's approach to mitigating motion sickness evolves as we continue to learn from user experiences in VR. In fact, recent research has shown that simultaneous presentation of sound and motion can help reduce motion sickness in VR (Yeo, Kwon, and Park 2022).

We are also figuring out new ways to integrate multiple modalities, such as sounds, visuals, and haptics, for immersive and interactive content delivery. In this book, we will discuss these emerging perspectives, exploring some of the current best practices in VR audio design within multimodal contexts. Nevertheless, seeing that these best practices are being identified as we speak, an audio designer who is just getting started in this field has ample room to innovate and play a decisive role in the evolution of VR. Although the learning curve for VR production can seem intimidating at first, the novelty of the medium implies that the expertise gap between seasoned and novice practitioners is much narrower than it is in more established creative domains, such as music, film, and games.

Over the next decade, there will be an increasing demand for immersive media specialists not only to meet the needs of a rapidly expanding medium but, more importantly, to help shape its future. As VR continues to mature into a new paradigm for media production, it will push us to reimagine our existing notions of content creation and consumption. We will face challenging yet compelling questions about the next generation of creative practices at the intersection of audio and VR. What is the role of sound in an increasingly interactive, immersive, and multimodal media landscape? How do we compose, perform, and produce music for and within VR? How do audiences engage with immersive content? How will the collaboration of disciplines and sensory modalities in VR elevate our artistic experiences moving forward? Throughout this book, we will address some of these questions while also posing new ones. We will take a hands-on approach to leveraging the unique possibilities of this medium as they manifest today. It is worth acknowledging that a guide that focuses on an emerging medium can be prone to obsolescence. To mitigate this, we will support our discussions with existing media theories and practices, while also making predictions about how VR technology might evolve and chart its own path in the future. We will explain VR audio concepts and techniques on a continuum that spans physics, perception, and technology. In doing so, we will formulate ideas that should be applicable to both current and future immersive media systems. While tomorrow's VR may look vastly different from today's, this is one of the most promising times in history to be curious about VR. It is this curiosity that lies at the core of this book.

Chapter 2

History of Immersive Audio

A Survey of Milestones and Techniques

While applications for immersive audio such as VR have started to reach a mainstream audience in the last few years, surround sound, especially for film, has been around for many decades. The use of surround sound in movie theaters largely drove the general public's experience of immersive audio, starting with Disney's *Fantasia* in the 1940s. Early versions of surround formats employed only a single surround audio channel, even if they were reproduced over multiple loudspeakers, whereas current immersive audio technologies can support hundreds of channels, allowing sound mixers greater control over sound source placement and therefore a much more immersive experience for the viewers. In this chapter, we explore the history and technical development of immersive audio and surround sound, also known as multichannel audio, up until today. We discuss early experiments in surround sound for musical applications, as well as those for film and games. We describe notable technologies such as the SoundField microphone, developed by Michael Gerzon and Peter Craven in the 1970s, along with the Ambisonics reproduction format, both of which were arguably ahead of their time. We also explore many of the multichannel formats that have been developed, concentrating mainly on developments since the mid-twentieth century.

2.1 Stereo and Beyond: Early Experiments in Surround Sound

VR invites users to be immersed in environments that are independent of their physical surroundings. Successful VR experiences require users to suspend their disbelief and adopt the perception that they are physically situated in a simulated world. To provide users with these kinds of experiences, engineers and designers create visual and auditory simulations consistent with the virtual world they wish to portray. While stereoscopic visual images coordinated with user movement create a sense of place, properly engineered 3D audio that responds dynamically to user movement is crucial to create a sense of depth and envelopment. Without accompanying 3D or spatial audio, VR experiences are significantly diminished.

As we look back to the history of spatial and immersive audio, we discover a long line of experiments and innovations aimed at increasing users' sense of envelopment and engagement with music or visual media. While early innovators in spatial audio may not have had VR applications in mind when they developed their inventions, their discoveries helped further our understanding of spatial audio, especially in terms of sound

DOI: 10.4324/9781003048022-2

localization, virtual room acoustics, distance perception, and listener envelopment, all of which is applicable to VR.

How did we arrive at our current understanding of 3D audio for VR? What were some of the important milestones in the development immersive audio technologies? In this chapter, we explore some of the highlights and important breakthroughs in the history of immersive audio for music and film that led to current technologies for sound spatialization.

Loudspeakers have long been the primary way to reproduce audio signals, and scientists and engineers have experimented with the number and arrangement of loudspeakers to understand and improve immersive audio systems. While modern VR systems primarily rely on binaural audio delivered over headphones, loudspeaker reproduction of spatial audio has played and continues to play an important role in the delivery of immersive experiences, particularly in venues like movie theaters and concert halls. Therefore, this chapter will introduce some highlights in the development of loudspeaker-based multichannel audio throughout the twentieth and twenty-first centuries.

As a technique to present spatial audio over headphones, binaural audio can provide a sense of location and depth using only two audio signals. These two signals, one for each ear, recreate the timing, level, and filtering of sounds that would have arrived acoustically at the left and right ears. Whether these two ear signals are recorded, transmitted live, or synthesized, binaural audio takes advantage of the mechanics of human hearing. The term "binaural audio" refers to a method of audio capture and delivery. This is distinct from "binaural hearing," a term that refers to our ability to perceive a three-dimensional auditory experience with only two ears. Although some of the earliest experiments in binaural audio date back to the late 1800s, contemporary audio researchers continue to refine the capabilities of binaural audio to deliver realistic 3D sound. Current binaural signal processing techniques are sophisticated and can create an astonishing sense of three-dimensional realism. We discuss issues related sound localization in Section 3.2, binaural audio and HRTFs in Section 3.4, sound reproduction over headphones in Section 4.2.1, and binaural rendering in Section 4.4.6.

In 1881, about four years after Thomas Edison filed a patent application for his invention of the first device to record and play back sound, one of the first demonstrations of binaural audio occurred at the International Exhibition of Electricity in Paris. Designed by Clément Ader, the "*Théâtrophone*" (literally, "theater phone") was a two-channel transmission system that used pairs of microphones positioned near the front of the stage at the Paris Grand Opera. Ader and his team wired each pair of microphones to a pair of telephone receivers (one for each ear) located in a separate building about 3 km (1.86 miles) away from the opera house. To allow up to eighty people to hear the Paris Opera performances at any one time, engineers set up twenty telephones with left and right earpieces in each of four separate rooms. An article published in the scientific journal *Nature* at the time described the experience of listening to the binaural telephone sound demonstration like this:

> The effect is in fact captivating; for the singing of the Opera is positively better heard in this way than in the Opera House itself; the words are more distinct, and the delicate turns of sound are better rendered.
>
> (Lockyer 1881)

In an article published in *Scientific American* that same year, another author described the Paris audio system's stereo image and its remarkable ability to produce a sense of space and allow listeners to track the singers' locations:

> ...in listening with both ears at the two telephones, the sound takes a special character of relief and localization which a single receiver cannot produce. As soon as the experiment commences the singers place themselves, in the mind of the listener, at a fixed distance, some to the right and others to the left. It is easy to follow their movements, and to indicate exactly, each time that they change their position, the imaginary distance at which they appear to be.
>
> (*Scientific American* 1881)

Looking back on this technology now, it is hard to imagine that the sound quality of this demonstration could have been anywhere near the sound quality we expect from current audio equipment. Yet the spatial effect and auditory perspective produced by a pair of microphones at the front of the stage was apparently quite striking to listeners at the time. Ader's work in this early form of stereophony was a demonstration of spatial audio and the powerful effect it can have on listeners, even when used with some of the earliest audio equipment available.

2.1.1 Blumlein and EMI

Any chapter on the history of immersive audio would be incomplete if it did not mention the work of British electronics engineer Alan Dower Blumlein. As a testament to the magnitude of Blumlein's impact on the world of audio engineering, audio engineer and former Vice President of CBS Labs Emil Torick wrote, "It has been said that the history of audio can be divided into two periods: *before Blumlein* and *after Blumlein*" (Torick 1998). Working at EMI in the United Kingdom in the 1930s, Blumlein made significant contributions to the development of stereo recording and reproduction. His inventions ranged from stereo microphone techniques to stereo reproduction on vinyl records. Not only are the stereo microphone techniques he invented in the 1930s are still being used today but the principles of his pioneering work on coincident stereo microphone techniques have been extended to current multichannel microphones and recording techniques. Ambisonics is one of the recording techniques that grew out of Blumlein's work and has been widely used to make immersive audio recordings for VR. We discuss Ambisonics microphones further in Section 4.3.3.

Blumlein's best known invention among recording engineers is probably his eponymous stereo microphone technique. It is perhaps less well known that he also invented the classic Mid/Side (M/S) stereo microphone technique. These two techniques were described in Blumlein's 1933 patent, reprinted in the *Journal of the Audio Engineering Society* in 1958. Both techniques specify a pair of coincident figure-8 (also known as bidirectional) microphones,[1] angled 90° apart. However, the two stereo techniques differ in their orientation and signal routing as we describe below (Blumlein 1958).

Recording engineers now commonly refer to only one of these two techniques as a "Blumlein stereo pair," while the other one is called a "Mid/Side stereo pair." The Blumlein stereo microphone technique, as we now refer to it, consists of two coincident figure-8 microphones with one angled 45° to the left and the other angled 45° to the

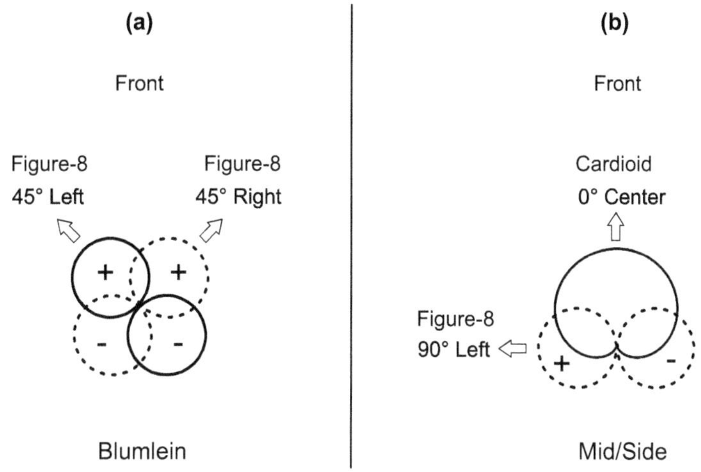

Figure 2.1 A drawing of microphone polar patterns for the Blumlein (a) and Mid/Side (b) stereo microphone techniques. Note the angle of orientation for each microphone.

right, giving a total acceptance angle of 90°. Due to the microphone orientations, sound arriving from directly in front of the pair is 45° off-axis for each microphone.

In the M/S technique, two figure-8 microphones are angled 90° apart, with one microphone (the "M" or Mid component) aimed straight ahead to the front, and the other microphone (the "S" or Side component) aimed at 90° to the left side. While Blumlein originally specified that the front-facing microphone in the M/S pair should be a figure-8 microphone, many recording engineers now use a cardioid polar pattern for the front-facing microphone, paired with a side-facing figure-8 microphone.

2.1.2 Fletcher and Bell Labs

Around the same time that Alan Blumlein was researching and inventing coincident stereo recording techniques in the United Kingdom, engineers in the United States were doing their own pioneering work with stereo. Most notably at the time was Bell Labs engineer Harvey Fletcher, who is best known by audio engineers for his contributions to our understanding of the frequency response curves of human hearing. Through extensive listening tests with human subjects, he developed the equal-loudness contours, commonly known as the Fletcher-Munson curves, which illustrate that human sensitivity to sound varies with frequency. Generally, we are least sensitive to low frequencies and high frequencies, and most sensitive to frequencies between about 3,000 Hz to 4,000 Hz. This is especially true at quiet listening levels. For example, for a listener to perceive a 50 Hz tone and 4,000 Hz tone to be the same loudness, the 50 Hz tone needs to be about 60 decibels (dB) louder than the 4,000 Hz tone, when reproduced at a quiet listening level (at the threshold of hearing). At a high listening level (say, around 100 decibels of sound pressure level (dB SPL)), the 50 Hz tone only needs to be about 8 dB louder to be perceived as the same loudness as the 4,000 Hz tone.

More relevant to the development of immersive audio, Fletcher is also generally regarded as the father of stereophonic sound, because he was among the first to demonstrate the spatial effect of sound. In 1916, he was hired by Western Electric Research Laboratories, which became Bell Telephone Laboratories in 1925. He eventually got promoted to Director of Acoustical Research at Bell Labs, a position he held until his retirement in 1949. Fletcher is said to have "single-handedly...created the fields of communication acoustics and speech and hearing as we know them today" (Allen 1996). He directed research that resulted in some of the first stereo recordings and the first live stereo transmission. He also led notable demonstrations of spatial audio during his tenure at Bell Labs.

At the 1932 World's Fair in Chicago, Fletcher and Bell Labs showcased binaural audio with a more sophisticated demonstration than Ader's *Théâtrophone* fifty years earlier. The main difference between the two demonstrations is that instead of simply using two spaced microphones, Fletcher used a mannequin called Oscar, with microphones fitted in its cheeks. Because he placed microphones on the sides of the mannequin's head, the microphone signals naturally contained interaural time and interaural level differences comparable to those of human hearing. For Fletcher's demonstration, the mannequin was placed on a glass-enclosed stage and numerous headsets were placed around the outside of the glass enclosure. Engineers fed the mannequin's microphone signals to listeners' headsets, with the left microphone routed to the left side of the headset and the right microphone routed to the right headset. For the demonstration, a person on stage walked around the mannequin and spoke. The glass enclosure prevented listeners from hearing the speech through the air, but they were able to hear it through headphones while watching the demonstration through the glass. Listeners reported being startled because they could hear someone walking around them. That's how powerful the binaural effect can be.

Collaboration between the arts and sciences has often led to important scientific and engineering breakthroughs and the field of spatial audio is no exception. In 1932, Fletcher made a stereophonic test recording of the Philadelphia Orchestra conducted by Leopold Stokowski. Stokowski was one of the most famous conductors of all time and happened to be intensely interested in multichannel and spatial audio. To demonstrate stereo recording, Fletcher and his team placed two microphones in front of the orchestra and connected their outputs to two styli cutting two separate grooves in a wax disc, one stylus starting from the outside and working in, and the other starting from the inside and working out.

In contrast to Blumlein, who emphasized coincident microphone techniques, Fletcher's Bell Labs colleagues J. C. Steinberg and W. B. Snow conducted experiments with multiple microphones spaced across a stage in a hall that were then reproduced by multiple loudspeakers in front of an audience in another space (Steinberg and Snow 1934). They theorized that an infinite number of microphones and an infinite number of loudspeakers would be ideal for reproduction of sound source azimuth and depth, a precursor to the development of wave field synthesis that would come fifty years later. Based on this theory, they experimented to identify a practical number of microphones and speakers and found that a three-channel arrangement provided an acceptable level of sound localization accuracy.

In 1933, Fletcher, Steinberg, and Snow demonstrated a three-channel stereo transmission of the Philadelphia Orchestra a technique they referred to as "auditory perspective." Signals from three microphones placed in front of the orchestra in the Academy of Music in Philadelphia were transmitted to Constitution Hall in Washington, D.C., about 224 km

(139 miles) away. These signals were then reproduced over three loudspeakers placed on stage in positions approximating the positions of the microphones in the Academy of Music in Philadelphia. The three-channel spatial audio reproduced in Washington created the illusion that the orchestra was performing just behind the stage curtain. Stokowski "manipulated the electric controls from a box in the rear of Constitution Hall while the orchestra, led by Associate Conductor [Alexander] Smallens, played in the Academy of Music in Philadelphia" (Fletcher 1934). A live audience in Washington experienced the three-channel audio reproduction of the orchestra. Fletcher, in a paper describing the demonstration, commented:

> Judging from the expression of those who heard this concert, the development of this system has opened many new possibilities for the reproduction and transmission of music that will create even a greater emotional appeal than that obtained when listening to the music coming directly from the orchestra through the air.
>
> (Fletcher 1934)

There was a sense, even during these early experiments, that spatial audio reproduction, could match or even surpass the experience of live music in a concert hall if done properly. Fletcher understood that music transmission and reproduction needed a system with wider bandwidth than was (and still is) typical of that used for speech transmission. Surprisingly, the entire system from microphones, transmission lines, power amplifiers, and loudspeakers was reported to have a bandwidth that ranged from 40 Hz to 15,000 Hz and a dynamic range of 70 dB (Affel, Chesnut, and Mills 1934).

2.1.3 Disney's Fantasound

Bell Labs was not the only organization engaged in significant audio engineering innovation in the United States in the 1930s and 1940s. Engineers at Walt Disney Studios were also at the forefront of audio technology development around that time. Their inventions included multichannel surround sound systems, simultaneous multitrack recording, close miking for orchestral recording and production, the pan pot, and the click track.

All these technologies, which audio engineers still use regularly today, were invented to support the company's release of the animated feature *Fantasia* in 1940. William E. Garity and John N. A. Hawkins, audio engineers at Walt Disney Studios in Burbank, California, invented a new multiple-speaker system, known as "Fantasound," which was used for the film. Walt Disney himself talked specifically about the importance of multichannel audio playback for *Fantasia* to give listeners a feeling of presence or "being there." He said,

> We know that music emerging from one speaker behind the screen sounds thin, tinkly, and strainy. We wanted to reproduce such beautiful masterpieces as Schubert's "Ave Maria," and Beethoven's *Sixth Symphony* so that audiences would feel as though they were standing on the podium with Stokowski.
>
> (Boone 1941)

Reaffirming his deep interest in the power of immersive audio and the importance of audio engineering to further the artistic goals of music performance, Leopold Stokowski was again involved in audio technology development, now with Disney as conductor of *Fantasia*'s musical score.

Fantasia's recorded soundtrack consisted of three tracks of audio, one for the center channel and another two that a mixing engineer panned around the theater manually. To implement this panning technique, Disney Engineers had to invent the panoramic potentiometer, more commonly known as the pan-pot. In collaboration with RCA, the engineers built and tested ten versions of Fantasound to determine the number of loudspeakers needed and their placements in the theater. The first version of Fantasound, called *Mark I*, involved three loudspeakers in the front and two loudspeakers in the rear, a channel configuration not unlike the 5.1-channel format that would eventually become somewhat of a standard for surround sound for movies starting in the 1990s. The *Mark II* system added a loudspeaker on each side wall and a single loudspeaker in the ceiling, which was remarkable for its time. Subsequent versions of Fantasound used a tone-operated gain adjusting device, called "Togad," which allowed panning to be automated the same way for each showing. The Fantasound system was integral to *Fantasia* and as such it was taken on a roadshow to movie theaters around the United States starting in 1940. Unfortunately, due to the expense and time involved to set up this elaborate sound system, it was installed in only about a dozen movie theaters before the roadshows ended in 1941. The film was far ahead of its time for its use of multichannel audio, but regrettably, only those who attended the roadshow at the time will ever know what it sounded like.

In a paper about sound for picture technology and its relation to live entertainment, Fantasound inventors Garity and Hawkins said that "motion picture entertainment can evolve far beyond the inherent limitations of live entertainment" (Garity and Hawkins 1941). Many decades prior to the current explosion of interest in VR, the audio engineers at Disney, who were responsible for the first surround sound system for film, envisioned that audio and video technology would enable us to go beyond what is possible with live performances. As Jesse Klapholz wrote, "Fantasound is technically as advanced today as the current Dolby cinema format. The innovations by RCA and Disney with Stokowski as a catalyst attest to the fusion of art, technology, and imagination" (Klapholz 1991).

2.1.4 The Philips Pavilion

Back on the other side of the Atlantic and almost two decades later, the Philips Pavilion provided another early example of immersive audio, but this time with experimental program material. Built at the 1958 World Exposition in Brussels, Belgium, the Philips Pavilion showcased composer Edgard Varèse's 8-minute electronic music piece, *Poème électronique* ("Electronic Poem"). The composition, generated from electronically synthesized and recorded sounds,[2] existed as a three-track tape recording that was spatialized over 350 loudspeakers embedded into the interior walls of the Pavilion.

Willem Tak, the acoustical consultant on the project, described the sonic goal of the system:

> The listeners were to have the illusion that various sound-sources were in motion around them, rising and falling, coming together and moving apart again, and moreover the space in which this took place was to seem at one instant to be narrow and "dry", and at another to seem like a cathedral.
>
> (Tak 1958)

The first track of the three-track tape contained the piece itself, while tracks two and three were reserved "specifically for reverberation and stereophonic effects" (Tak 1958). On a second tape, 15 tracks[3] of control signals determined audio signal routing to each of the various loudspeakers to create the impression of sounds moving around the space. Each of the 15 tracks contained 12 control signals each with a fixed amplitude and fixed frequency, which gave them 180 (15 times 12) control channels. The 12 control signals on each track, which ranged in frequency from 900 Hz to 10,500 Hz, were used to switch loudspeakers on and off and to control the lighting effects. Special circuits were tuned to each control signal frequency so that a specific loudspeaker amplifier was turned only when the frequency to which the circuit was tuned was present. The control signals were recorded with specific timings to ensure that each of the three tracks in Varèse's piece was reproduced by the intended loudspeaker at the intended time. Despite only three tracks of actual audio, Philips engineers created the perception of sound moving through the space by controlling which of the 350 loudspeakers were reproducing the sound at any moment in the piece (de Bruin 1958).

Projected images, chosen by the Pavilion's architect Le Corbusier, accompanied *Poème électronique*. The resulting multimedia experience was described as "the first electronic-spatial environment to combine architecture, film, light and music to a total experience made to function in time and space" (Lopez 2011). The Philips Pavilion was demolished shortly after the fair ended, but within the six-month period that it was open, about two million people experienced this cutting-edge experimental immersive audio environment.

2.1.5 Quadraphonic Sound

Quadraphonic sound, commonly called "quad," was designed to use four channels of audio. The format began to generate interest in the 1960s and gained popularity in the early 1970s as an alternative to stereo since it offered surround audio playback for music recordings. The preferred quad loudspeaker arrangement included two front channels like traditional stereo along with two mirroring rear channels, as shown in Figure 2.2. The four channels formed a square. In current nomenclature we would refer to quad as a 4.0-channel or 2/2 channel surround system. This loudspeaker placement had the advantage that stereo recordings could be played back through the front loudspeakers of a quad system without the need to rearrange the loudspeakers.

In 1960, not long after the Philips Pavilion closed, composer Karlheinz Stockhausen completed his piece *Kontakte* in Cologne, Germany. One of his most famous works, it was groundbreaking not only in its use of electronically generated sounds but also in the spatialization of those sounds. Four channels of audio fed four loudspeakers, located at the corners of a square, which reproduced the electronic sounds around the audience. There were two versions of the piece. Both employed a four-channel tape recording of electronic sounds, but one version added live percussion and piano.

In 1969, about a decade after the multichannel demonstration at the Philips Pavilion, American musician and inventor Peter Scheiber introduced his new audio matrix for four-channel surround sound (Eargle 1976). Scheiber's invention helped set off a decade-long interest in quadraphonic sound for music listening. It allowed vinyl records, the primary music distribution format back then, to deliver quad sound to the masses. Scheiber's audio matrix, known as a 4-2-4 matrix, allowed four channels of audio to be matrixed or mixed into two channels for distribution. In this technique, a quad encoder

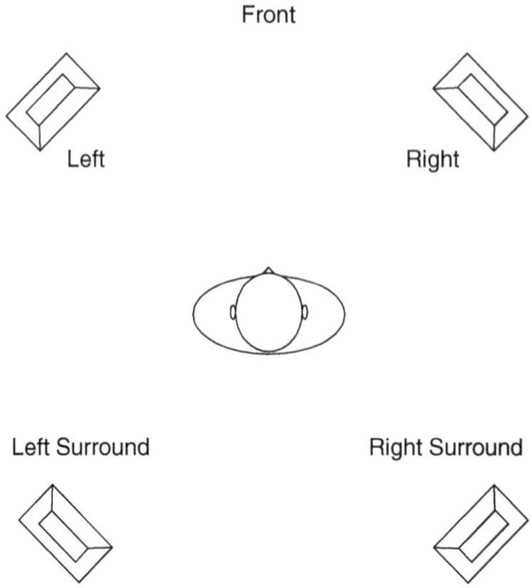

Figure 2.2 A quadraphonic or "quad" 4-channel loudspeaker layout with two front loudspeakers and two rear loudspeakers.

matrix applies specific amounts of gain and phase shift to each of the four channels as they are mixed to two channels. Two-channel encoded mixes are then "de-mixed" or up-mixed back to four channels during playback. This matrixing technique was essential since vinyl records could only store two channels of audio.

Many music recordings were released on vinyl records in quad during the 1970s across a variety of competing quad audio formats with dedicated hardware matrix decoders. CBS/Columbia's SQ matrix (for stereo-quad) and Sansui's QS matrix (for quadraphonic synthesizer) were the main competing formats in the 4-2-4 matrix market, and both were based on Scheiber's work. According to Eargle, the Sansui QS Matrix has essentially the same encoding and decoding coefficients as the Scheiber matrix, but has a more natural sound than the Scheiber format because the QS matrix ensures that the rear channels are never more than 90° out of phase (Eargle 1976). Despite the similar names, the QS and SQ matrix formats were not compatible with each other, which meant that listeners had to match the matrix decoder to the encoded format for playback.

In addition to 4-2-4 matrix formats, other quad formats supported four discrete or independent channels by encoding difference signals for multichannel audio in the ultrasonic region above 20 kHz. One such format was the Compatible Discrete 4, or CD-4 (not to be confused with the compact disc), a discrete four-channel system developed by Japanese electronics company JVC in 1971 and adopted in America by RCA where it was known as RCA Quadradisc (Brice 2019). The main issue with any format that uses ultrasonic frequencies on vinyl records is that it relies on the high frequency response of the vinyl to remain intact after repeated plays. The problem is that each time a vinyl record is played, its grooves wear down, resulting in high frequency loss. Thus, CD-4 was a commercial failure.

One issue with quadraphonic sound reproduction systems from a sound quality point of view, regardless of the encoding/decoding format, is the instability of phantom images on the sides and to the rear. Due to the way we localize sound, standard amplitude panning between two loudspeakers on one side (e.g., the left front and left rear loudspeakers) is problematic for accurate placement of lateral phantom images. Unlike the stable phantom images we can generate between a frontal pair of loudspeaker, the perceived phantom image between two side speakers is not located at the midpoint between them when they produce the same signal at equal sound levels. There is also a lot of variability between listeners in terms of lateral sound localization with two side loudspeakers. Furthermore, the perceived location of a side phantom image changes with small movements in listening position, as we know from multichannel audio localization research (Martin et al. 1999a; Theile and Plenge 1977).

Ultimately quad's allure faded by the end of the 1970s for several possible reasons. For one, the various proprietary formats were confusing to consumers. It's also possible that they may not have wanted to install two additional speakers in their living rooms. At the same time, record producers had difficulty taking full advantage of the format which resulted in disappointing mixes. Perhaps the main reason quad died was because record labels could not make a profit from the format so they stopped producing quad recordings (Torick 1998).

While quad fell out of favor by about 1979, there has been some renewed interest in quad matrixing with the release of a freely available DAW plugin called QUARK, for Quadraphonic Universally Accessible Resource Kit (available at quark.cykik.com). The QUARK plugin is based on the 1970s Sansui QS matrix format and the plugin developer has released music mixes in quad since 2018 starting with Suzanne Ciani's album *LIVE Quadrophonic*.

2.2 Surround Sound for Theaters and Broadcast

Robert Fine is probably best known for the Mercury Living Presence[4] classical music recordings he engineered with record producer Wilma Cozart Fine. In the mid-1950s, Robert Fine invented a technology that allowed single-channel film soundtracks to produce a directional sound system over three front cinema loudspeakers (Fine 1954). Called "Perspecta Stereophonic Sound," it smoothly steered a film's single, mono audio channel between the left, center, and right loudspeakers using low-level, low-frequency control sine tones that were mixed with the film soundtrack. With a specific control frequency assigned to each loudspeaker—30 Hz for left, 35 Hz for center, and 40 Hz for right—the amplitude of the tones determined the audio signal amplitude in the respective loudspeakers. The Perspecta Stereo Integrator hardware, which was manufactured by the Fairchild Recording Equipment Company, could detect the low-level control tones, use them to control the location of the sound, and then filter out the low frequency tones before the soundtrack audio reached the loudspeakers. Since it merely panned a mono signal across three loudspeakers, Perspecta was not a true stereophonic or surround system. Nonetheless, it represented a significant interest in moving beyond mono audio. It also retained mono compatibility since films that included the Perspecta control signals could be played back on a film projector without the Perspecta Integrator hardware. Perspecta was supported by MGM Studios and Paramount Pictures, who used it on nearly every film release for about four years in the mid-1950s.

Despite the advanced surround sound capabilities of Fantasound in the 1930s and '40s, commercial movie theaters typically had only one loudspeaker until the mid-1970s. It was placed behind the center of the projection screen, which is perforated to allow sound transmission to the audience. While a mono loudspeaker may not convey much of a sense of immersion or envelopment, its central location does the important job of locking dialog to the center of the screen. Two-channel stereo (left and right) audio playback without a center channel, as is typical for music recordings, is problematic for movie sound playback in a theater mainly because of phantom image localization and its dependence on listener location. If a theater has only two widely spaced loudspeakers (positioned at the left and right sides of the screen) without a center channel, audiences will localize any center-panned audio, such as dialog, to the closest loudspeaker. Only those seated in the center of the theater, that is, equal distance from the left and right speakers, perceive the most accurate phantom images. Most people in a theater are off center, therefore localizing the dialog to one side of the screen. The resulting location mismatch between actors and their voice has an unintended consequence of drawing viewers' attention to this mismatch, which degrades the movie experience.

This raises a broader point to keep in mind for VR production: anything that causes viewers to be aware of the technology, such as audiovisual timing or synchronization issues, audiovisual localization mismatches, or mismatches between the visual representation of the size of a space and the acoustical modeling, reverberation, and reflections, can disturb the VR experience. It is like live sound reinforcement: the average audience member or VR user (without any training in audio engineering) pays attention to the sound only if there is an issue with it. If the audio is problem-free, no one (except the audio engineers!) pays attention to the audio quality itself. Instead, users pay attention to the VR experience, the music, the movie, or the game itself.

2.2.1 2D Surround Sound

Since most movie theaters had only a single center loudspeaker until the mid-1970s, film soundtracks were recorded as a single optical track on 35 mm film until there was more substantial interest in surround sound movie soundtracks. Going from mono to surround sound meant solving two non-trivial technical issues. One was to encode surround sound (that is, multiple channels of audio) on film, which at the time contained only one channel of audio. The other issue was to reduce the extra noise produced by having more than one optical audio track on film.

2.2.1.1 Dolby Stereo

With its optical audio tracks, film became far too noisy when a second channel of audio was added. The noise level increase occurred because the available space on the film, originally reserved for a single mono optical track, had to be split in two. The reduction in track width caused an increase in noise level, therefore reducing the signal-to-noise ratio. To mitigate this, Dolby applied noise reduction to the film based on a technique similar to a process invented at Bell Labs in 1932 (Dienstfrey 2016). In 1976, they released Dolby SVA (stereo variable area), which used two-channel optical tracks on 35 mm film and incorporated this noise reduction technology. In addition to noise reduction, Dolby SVA added four-channel surround sound using only two recorded tracks of audio. Dolby SVA was released alongside a separate format for six-track magnetic 70 mm film, known as Dolby Six-Track, under the common brand name Dolby Stereo.

After noise reduction enabled two optical tracks on film, the next issue to solve was to encode four channels of audio into the two optical tracks. Engineers knew from developments in quadraphonic sound that it was possible to produce four-channel surround audio from only two tracks of audio using the 4-2-4 matrix technology Scheiber invented several years earlier. Scheiber licensed his quad 4-2-4 matrix technology to Dolby for its cinema and consumer matrix systems (Scheiber 2003). The 4-2-4 matrix technology they incorporated into Dolby Stereo had previously been used primarily in quad for music applications, but Dolby chose not to position the four channels equally spaced around the room as the quad format had specified. Instead, since they were focused on sound for movies, a front center channel was required to lock the dialog audio image to the center of the screen. Because of the inclusion of a center channel in Dolby Stereo and the way that the four channels were combined in the 4-2-4 matrix, a Dolby Stereo soundtrack was still compatible with those theaters that had not upgraded their sound systems by that point and still had only a single mono channel sound system. The first movie to feature a Dolby Stereo soundtrack was *A Star Is Born*, released in 1976.

It may be worth a note here that the number of "channels" does not necessarily correspond to the number of loudspeakers. For example, while the left, center, and right audio channels each correspond to specific left, center, and right loudspeakers located behind the projection screen at the front, movie theaters often have several surround loudspeakers arranged across the side and back walls. In the case of a four-channel mix with a single surround channel, such as Dolby Stereo, all surround loudspeakers reproduce the same channel of audio. Therefore, in the Dolby Stereo format, the surround loudspeakers produce a monophonic image. While a single surround channel produces a less immersive sound field than two or more surround channels, Dolby Stereo with its four channels of audio was nonetheless a huge improvement over just a single mono center loudspeaker for movie soundtracks.

While a 4-2-4 matrix allows a much more immersive film soundtrack, it is not a perfect system. The main drawback of a matrix encoding/decoding process such as 4-2-4 is that it reduces channel separation because the matrixing process does not maintain four discrete channels. While 4-2-4 matrix systems brought more immersive sound to more movie-going audiences, the technology was not universally admired. In a critique of the technology, audio engineer and writer John Eargle pointed out that due to their limited channel separation, matrix systems might be considered as simply stereo enhancement systems (Eargle 1976).

2.2.1.2 *Dolby Surround and Dolby Pro Logic*

The Dolby 4-2-4 matrix went through several improvements and iterations over many years. Dolby Stereo, the name of the technology used in movie theaters, was simplified and adapted for use in home systems under the name Dolby Surround in the early 1980s. The consumer-focused Dolby Surround technology likewise used a passive decoding process to derive a surround channel from the left and right channels. Passive decoding simply mixes the Lt and Rt channels to extract the center channel. As we saw above, surround channel information is canceled when Lt and Rt are summed. To extract the surround channel, the decoder subtracts the Lt and Rt channels which causes the 180°-out-of-phase surround channel signal level to double. Subtracting the Lt and Rt channels also has the effect of canceling any information which may be in phase between the two channels such as the center channel audio.

The problem with a passive surround decoder like Dolby Surround is the limited channel separation, which, as we mentioned above, can be as low as 3 dB. In practice this means that any sound panned to the left channel, for example, will also be reproduced by the surround channel, although at a slightly lower level. To help improve channel separation at least for one-at-a-time signals, some 4-2-4 matrices employ an active gain-riding circuit called a "logic" that detects prominent signals in any one channel and attenuates the overall level in adjacent channels. It does this by mixing an opposite polarity version of the channel audio which destructively interferes with the signal and cancels it. One of the problems with the gain riding logic is that it produces audible "pumping" due to the level changes in signals that otherwise would have had continuous levels. In 1987, Dolby replaced Dolby Surround with a new technology called Dolby Pro Logic (presumably so-named because it used a "logic" gain-riding circuit) for the consumer market, which was an active decoder like the one found in commercial cinemas called Dolby Stereo. A subsequent version of the technology, Dolby Pro Logic II, allowed listeners to take any two-channel stereo signal, regardless of whether it was originally a four-channel signal, and play it back across three (L, C, R), four (L, C, R, S), or five (L, C, R, SL, SR) channels. This technology expanded the number of channels from four-channel surround to five-channel surround, providing two channels for surround, and added a low frequency effects (LFE) channel, designated as a 0.1. Further development led to Pro Logic IIx, which allowed two-channel or 5.1-channel recordings to be up-mixed into 6.1- and 7.1-channel playback systems.

2.2.1.3 Discrete 5.1-Channel Audio and Dolby Digital

As mentioned above, the main problem with matrix surround processes is that channel separation is much less than optimal because channels must be combined and then decoded in the analog domain. Complete channel separation and independence is simply not possible in a matrix surround encoding-decoding process such as 4-2-4. To maintain channel separation, a surround system needs to support discrete channels, that is, channels that are completely independent and not combined or matrixed at any stage from capture to storage to reproduction. To support discrete multichannel systems Dolby developed Dolby Digital, also known as AC-3. Dolby Digital was formally launched in 1992 with the release of the movie *Batman Returns* (Todd et al. 1994) and it became one of the most widely-used movie soundtrack formats.

Dolby Digital AC-3 is a multichannel audio coding standard that supports up to 5.1-channels of audio. These independent or discrete channels include left (L), center (C), right (R), left surround (LS), and right surround (RS) channels, as well as a ".1" LFE (or subwoofer) channel with a frequency range of about 20 to 120 Hz. The format also provides stereo support automatically. The main drawback of AC-3 is its reduced audio quality due to the limited bit rate of the perceptually encoded or compressed audio data. It has a maximum bit rate of 640 kbit/s total to encode all six channels. For comparison, the bit rate of six uncompressed audio channels at a sampling rate of 44,100 Hz and 16 bits per sample is 4,233.6 kbits/s. In other words, the AC-3 encoder provides nearly a sevenfold reduction in data. Dolby managed to achieve market dominance in surround audio for movies and TV with AC-3 because every DVD (digital video disc) worldwide and all HD broadcasts in the United States use Dolby Digital ("Dolby Digital 5.1," n.d.).

Prior to digital projection, all movies were released on film, which meant that Dolby Digital audio data was placed between the sprocket holes of the film as an optical track. Since it is a digital signal, it is immune to noise issues that plague analog film sound that uses optical tracks. In commercial movie theaters with two surround channels (such as the 5.1 format), audio systems typically contain several surround loudspeakers on the left side and left back wall for the left surround (LS) channel and several loudspeakers on the right side and right back wall for the right surround (RS) channels. In cinemas, it is common to feed several surround loudspeakers with a single channel of audio since precise localization in the rear is not as crucial and having more loudspeakers in the rear helps to generate a better sense of immersion while maintaining even coverage of the seating area.

The most common 2D loudspeaker configurations include 5.1-, 7.1-, and 9.1-channel systems. Historically the 5.1-channel configuration has been the most common one. Around the same time that Dolby Digital was released publicly, the International Telecommunication Union (ITU) issued a recommendation in 1992 for a five-channel loudspeaker layout, known as 5.1-channel surround or 3/2 channel format, shown in Figure 2.3(a). The ITU recommendation, known by the designation ITU-R BS.775, was aimed at television broadcasting applications which requires stable front sound image, using three front loudspeakers—left, center, and right—along with two surround loudspeakers to provide a sense of spaciousness, and an optional LFE channel. The 7.1-channel audio system is an expansion of the 5.1-channel system and includes side speakers as shown in Figure 2.3(b). The inclusion of side channels in the 7.1-channel system helps

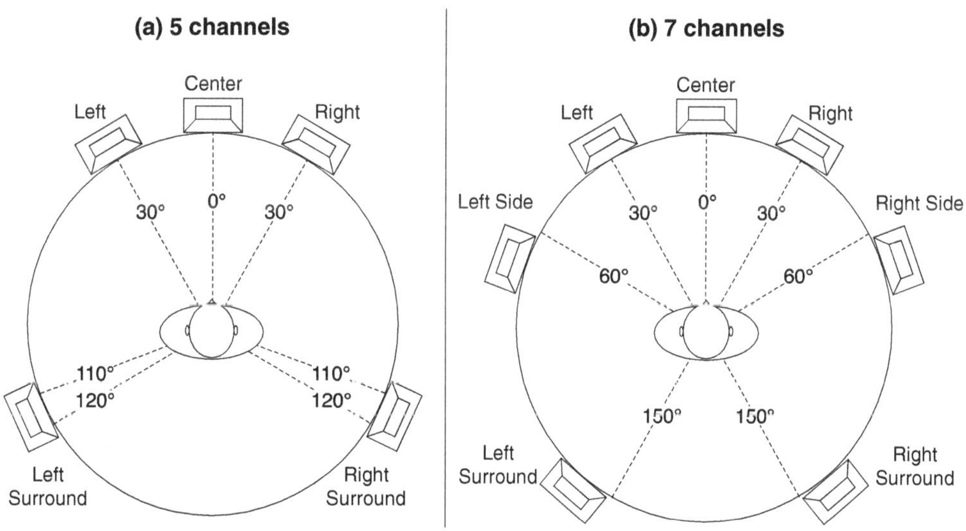

Figure 2.3 (a) The standard 5-channel loudspeaker layout recommendation, also known as 3/2 since there are three front loudspeakers and two surround loudspeakers, and (b) the 7-channel loudspeaker layout recommendation, also known as 3/4 since there are three front loudspeakers and four side and surround loudspeakers. The 7-channel recommendation states that the side and surround loudspeakers should be placed at equal intervals on the arc which measures from 60° to 150° from the center front reference. When a subwoofer or low frequency effects (LFE) channel is added, we refer to the loudspeaker layouts as 5.1-channel and 7.1-channel systems, respectively. The loudspeaker layouts are specified in the ITU-R BS.775 recommendation. Adapted from International Telecommunication Union (2022).

reduce localization errors, which can be especially problematic for side images in a 5.1-channel system. Side channels also help create continuity of sound from front to back.

2.2.1.4 Other Formats

DTS, the main competitor to Dolby Digital, offered their own multichannel perceptual encoder with arguably better audio quality due to higher supported bit rate for the encoded audio. The DTS perceptual encoder used less data compression than Dolby Digital and supported a bit rate of up to 1,500 kbit/s for six channels of audio, more than double the data rate of Dolby Digital. The main operational difference was that 5.1-channel DTS film soundtracks were stored on CD-ROMs (a data-only version of the compact disc format, since audio CDs do not support more than two channels of audio) that had to be synchronized with their respective films during playback. The requirement to synchronize two separate media (film and CD-ROM) made DTS more cumbersome to use than Dolby Digital and it was not adopted as widely.

Sony introduced a competing film surround audio format in the 1990s called Sony Dynamic Digital Sound (SDDS). Sony's format supported eight channels of audio: five front channels, two surround channels, and a low-frequency subwoofer channel. Sony specified five front channels, left, left-center, center, right-center, and right, for more even audio coverage in cinemas with large screens. The encoded SDDS audio was placed on the outer edges of a film. Supporting a bit rate of 2,200 kbit/s, SDDS also included redundant copies of audio tracks in case edges of the film got damaged and therefore degraded the audio data.

While cinema sound systems expanded the number of channels available, a few companies were aiming to bring 3D audio to consumers' music listening experiences. Several companies in the 1990s offered much simpler, albeit less effective solutions for three-dimensional audio reproduction over just two loudspeakers. Companies including QSound, Roland, SRS Labs, and Spatializer Audio Labs offered stereo enhancement processing and virtualizers for music playback over home audio systems. These stereo enhancement hardware devices and software algorithms process stereo recordings to produce the perception of 3D sound using only stereo loudspeakers using spectral binaural filtering and crosstalk cancellation. To achieve the strongest effect, listeners are required to be positioned in a very specific listening location, the sweet spot, to form an equilateral triangle with the speakers. While most of these plugins and hardware processors are not being produced anymore, the current Apple Music/iTunes player continues to offer a "Sound Enhancer" feature on laptops and desktop machines that seems to widen the stereo image and provide some equalization. Stereo enhancement processors operate within the limitations of two-channel loudspeaker reproduction systems and therefore do not offer the accuracy of multichannel audio systems.

2.2.2 3D Surround Sound

In natural hearing, acoustic sounds may be located at any elevation, so it makes sense for audio systems intended for 3D immersive audio reproduction to include loudspeakers above and below the listener. While the term *surround sound* has been used for decades, especially with regards to movie soundtracks, loudspeaker placement for this format has traditionally been limited to the 2D horizontal plane, that is, without overhead loudspeakers or height channels. Horizontal-only loudspeaker layouts are also known as *pantophonic* systems, which consist of only ear-level loudspeakers. Loudspeaker arrays that include height channels are known as *periphonic* systems, with loudspeakers above and possibly below the listener in addition to those at ear level. Some of the notable periphonic multichannel loudspeaker layouts include Ambisonics, Auro-3D, Dolby Atmos, DTS-X, Holman 10.2, NHK 22.2, and Sony 360 Reality Audio, which we describe briefly below.

2.2.2.1 Dolby Atmos

In 2012, Dolby introduced the Atmos format with the release of Disney/Pixar's movie *Brave*, the first feature film to be mixed entirely in this format. While 5.1-channel and 7.1-channel loudspeaker layouts continue to form the basis for Dolby Atmos, two main factors differentiate it from the company's earlier surround sound formats: overhead speakers and audio objects. Although overhead speakers are not required for Atmos, Dolby strongly recommends them for mix rooms. The format supports a range of loudspeaker layouts such as those that incorporate only ear-level loudspeakers, including 2.1-channel stereo, and 5.1-channel, 7.1-channel, and 9.1-channel surround. The most common speaker configuration in music-focused mix rooms is 7.1.4 channels, that is, seven ear-level channels, an LFE channel, and four overhead channels, as shown in Figure 2.4.

Movie theaters with Atmos systems can install up to 64 individually addressable loudspeakers, whereas home theaters can have up to 34 individual channels. The second differentiating factor from earlier surround formats is that Atmos provides support for what are called audio objects, which allow a mix to be played back on almost any loudspeaker configuration. Audio objects are simply audio tracks with accompanying metadata that specify 3D positional information, that is, the X, Y, and Z coordinates for each sound source in a mix. We discuss audio objects further in Section 4.4.2. The metadata that accompanies the audio objects allows all the sound sources to be rendered correctly during playback. The word "metadata" is simply information or data about the digital audio data. Audio tracks are often embedded with metadata such as song title, artist, composer or songwriter, copyright information, and so on. In the case of surround sound with audio objects, metadata provides sound panning and mixing instructions that an audio playback system uses to position sound sources according to the loudspeaker arrangement in use. Therefore, an Atmos mix can be translated to different loudspeaker configurations since a movie theatre sound processor can use metadata to position audio signals for the theatre's specific loudspeaker layout. Dolby Atmos mixes can also be converted to binaural signals for listening over headphones, but the format currently supports only sound sources positioned at ear-level and above, which can be problematic for VR.

2.2.2.2 NHK 22.2-Channel Audio

Pushing the channel count even higher to accompany ultrahigh-definition television (UHDTV) broadcasts, Kimio Hamasaki and his colleagues at NHK (Japan Broadcasting Corporation) first demonstrated a 22.2-multichannel audio system in 2005. This system specifies nine upper layer (height) channels, ten middle later channels (ear level), three lower layer channels in the front, and two LFE channels. Figure 2.5 shows the loudspeaker layout for the NHK 22.2 system.

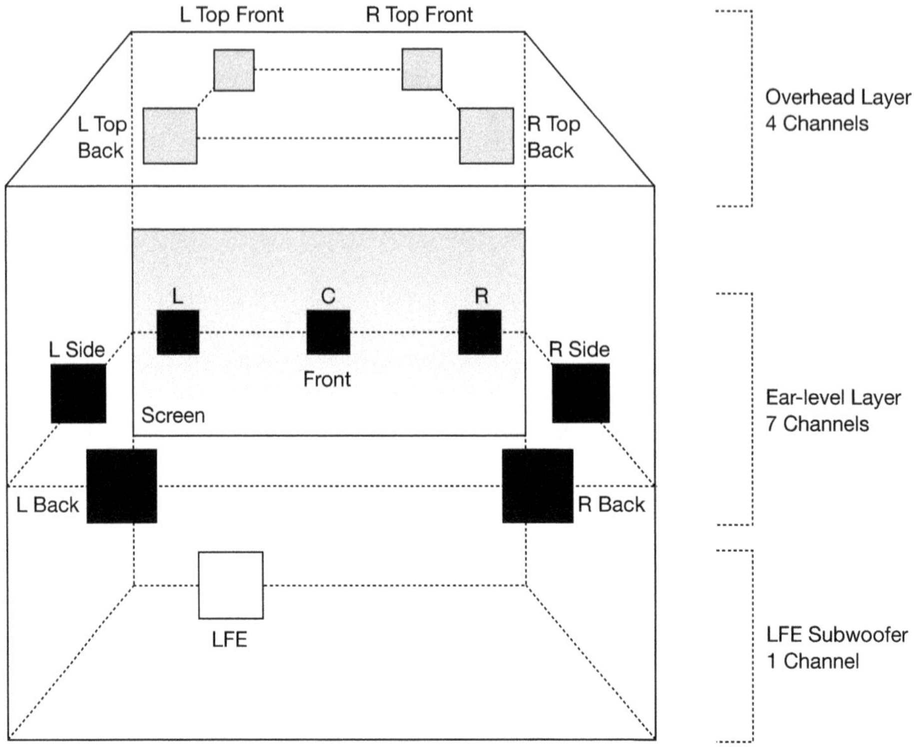

Figure 2.4 A diagram of the Dolby Atmos 7.1.4-channel loudspeaker layout that includes seven ear-level loudspeakers (three front, two side, and two back), one low-frequency effects (LFE) subwoofer, and four overhead loudspeakers. Each loudspeaker has an optimal angle or angle range: Left 22° to 30°, Center 0°, Right 22° to 30°, Left Side 90° to 110°, Right Side 90° to 110°, Left Surround 135° to 150°, and Right Surround 135° to 150°; Overhead Left Top Front 30° to 60°, Overhead Right Top Front 30° to 60°, Overhead Left Top Rear 120° to 150°, and Overhead Right Top Rear 120° to 150°. Overhead loudspeakers should be placed so that they have an elevation between 30° and 55° relative to the listening position. The side-to-side separation of the overhead loudspeakers should be 0.5 to 0.7 of the width of the overall loudspeaker layout.

From their listening tests, they found that the 22.2 system can produce better spatial audio quality than 5.1 multichannel audio. Listeners rated a higher degree of presence over a wider listening area with 22.2 than with 2.0-stereo or 5.1-multichannel, indicating a larger sweet spot. They found that 22.2-multichannel audio could provide a better impression of depth, naturalness, distance, listener envelopment when compared to 5.1-channel audio, and significantly better naturalness and listener envelopment than 10.2-channel systems (Hamasaki et al. 2004; Shim et al. 2010; Hamasaki et al. 2008). They proposed the following uses for each of the three loudspeaker layers:

- Upper layer: reverberation, sounds originating from above, such as airplanes
- Middle layer: the basic sound field, reproduction of envelopment
- Lower layer: water sounds, scenes on the ground when the picture presents a bird's eye view

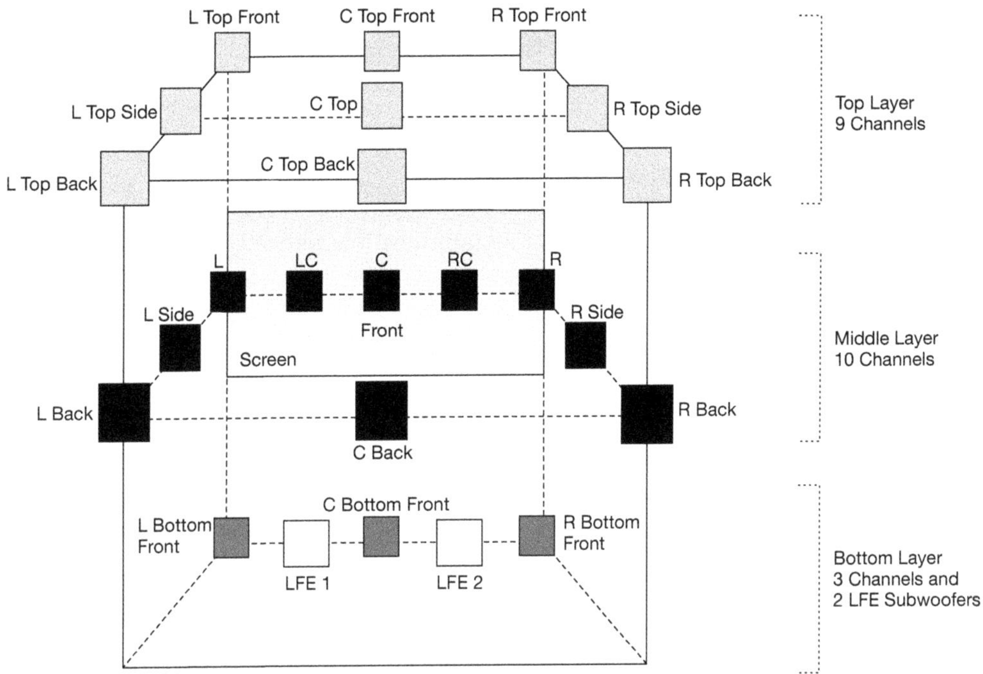

Figure 2.5 A diagram of the NHK 22.2 loudspeaker layout. Loudspeakers are located above, below, and around the listening area indicating a clear goal to produce immersive audio. The loudspeaker array includes nine upper (top layer) loudspeakers, 10 ear-level (middle layer) loudspeakers, three bottom layer loudspeakers below the screen, and two low frequency effects (LFE) subwoofer channels. Adapted from ITU-R (2012).

The NHK 22.2 system has primarily been an immersive audio research environment with some permanent installations at NHK, Fraunhofer IIS, and McGill University. NHK originally devised the system for broadcasting 22.2 audio along with UHDTV. In 2018, NHK began broadcasting 8K video along with 22.2 multichannel audio.

2.2.2.3 *Other Formats*

As an extension to the ITU 5.1-channel configuration, Tomlinson Holman, the creator of THX, and Chris Kyriakakis at the University of Southern California developed a 10.2-channel system to help reduce localization errors but placing channels in gaps of the 5.1 system (Holman 2000). It also introduces height channels. The system doubles the 5.1 system channel count by adding two wide front channels (60° L and 60° R), a center back channel between the surround channels, two widely spaced height channels (45° L and 45° R at an elevation of 45°), and a second LFE channel (making it 0.2). While this system has been demonstrated with great success at professional meetings for audio engineers and enthusiasts, it has not been adopted by movie theaters for the public.

In 2011, the movie *Red Tails* became the first to feature immersive sound using Auro-3D's 11.1-channel format. Auro-3D is a periphonic immersive multichannel format for cinema that includes three layers, but its layers are categorized differently than those in the NHK 22.2 system. Although there are no lower channels, the elevated and overhead speakers are treated as separate layers. The three Auro 3D loudspeaker layers include:

- Layer 1: ear-level surround, 0° elevation
- Layer 2: height or upper, 30° elevation
- Layer 3: top or overhead, 90° elevation

Auro-3D supports several loudspeaker configurations but generally maintains compatibility with 5.1-channel or 7.1-channel soundtracks. For home cinemas, Auro-3D supports several different loudspeaker configurations from Auro 8.0, which is mainly recommended for music listening since there is no center speaker (a channel that is usually prioritized for dialog in movies), up to Auro 13.1. Auro Technologies states that Auro 9.1 is the most efficient two-layer loudspeaker configuration that is compatible with 5.1-channel audio. Auro 9.1 employs five loudspeakers at ear level (L, C, R, LS, and RS) and four loudspeakers in the height position (L, R, LS, RS), plus a subwoofer channel (.1). As the most efficient three-layer loudspeaker setup compatible with 5.1-channels audio, Auro 10.1 adds a single overhead channel. The two main Auro-3D channel configurations for commercial cinemas are Auro 11.1 and Auro 13.1. Auro 11.1 employs five channels at ear level (layer 1), five height channels (layer 2) at 30° elevation aligned vertically with layer 1 channels, one overhead channel, and a subwoofer channel. Auro 13.1 employs seven channels at ear level (layer 1) to be compatible with 7.1-channel reproduction, five height channels (layer 2) at 30° elevation (like Auro 11.1), one overhead channel, and a subwoofer channel.

In 2015, DTS Inc. introduced the multichannel audio codec DTS-X, first for home theaters, then later for commercial cinemas. DTS-X is the main competitor to Dolby Atmos and likewise supports object-based audio and multichannel audio with height. DTS-X supports up to 11.2 channels, unlimited sound objects, and up to 32 loudspeaker locations.

Sony 360 Reality Audio specifies a 13-channel loudspeaker setup which includes a 5-channel horizontal array in the standard ITU 5-channel loudspeaker arrangement (i.e., 30° left, 0° center, 30° right, 110° left surround, and 110° right surround), a 5-channel height array with 30° elevation aligned vertically with the horizontal array (similar to the Auro 11.1 layer 2), and three bottom or lower speakers positioned 20° below, aligned vertically with the left, center, and right channels. The format does not specify an LFE channel. Sony offers a plugin called 360 WalkMix Creator to allow anyone to mix for the format and it includes binaural rendering for mixing on headphones.

2.2.3 *Continuing Work with Binaural Audio*

Although Harvey Fletcher was experimenting with dummy head recording in the 1930s, it wasn't until 1973 that microphone company Neumann first introduced a commercially available dummy head for recording. Currently sold as the Neumann KU 100 and dubbed "Fritz," it is modeled after an average sized adult human head and includes omnidirectional microphones built into its artificial ears. More recently a company called 3Dio began producing a binaural microphone at a much lower price than other dummy heads. Their model dispenses with the "head" and includes only two external ears (pinnae) with microphones embedded in them, spaced apart by about 15 cm (6 in.), equal to the average width of a human head. Recording engineers continue to find the dummy head to be a valuable tool for creating immersive recordings.

Some music streaming services now provide binaural music recordings, calling the format "spatial audio" rather than "binaural audio." These recordings are usually encoded in one of the two most used formats, Dolby Atmos or Sony 360 Reality Audio. The spatial audio versions support not only horizontal plane surround sound but also height information. Spatial audio formats like Dolby Atmos and Sony 360 Reality Audio support 3D immersive sound for multichannel playback, and importantly, since most people listen to music using headphones, the formats can render multichannel mixes for binaural playback over headphones. While binaural audio for music was once a niche area of interest with few binaural recordings available, it has now become a mainstream format as streaming services promote the mixing and remixing of recordings for spatial audio. In Sections 3.2 and 3.4, we discuss the perceptual underpinnings of sound localization and binaural hearing. Then in Sections 4.2.1 and 4.4.6, we offer technical details of binaural sound as a spatial audio format, highlighting its essential role in VR audio.

2.3 Soundfield Techniques

Surround sound often draws from or is built upon stereo techniques. Amplitude panning is used to position sound sources around a listening area whether two or more channels are used. Surround sound microphone techniques are often developed from the principles used in stereo microphone techniques. Neither stereo nor surround sound systems recreate a sound wave as it originally occurred; it's an approximation that creates an illusion for listeners. For example, two stereo loudspeakers that emit the same signal, let's say a recording of someone talking, produce a single phantom image in the center of the speakers. While the original sound was produced by a single sound source (the person talking), the phantom image of the sound source is produced with two sound sources (the two loudspeakers). Sound from the stereo loudspeakers sums at the listeners ears to create a convincing phantom image illusion of a sound source located between the loudspeakers. While the sound of the talker heard acoustically and the sound of the talker reproduced over loudspeakers as a phantom image may sound similar, the sound waves reaching a listener's ears in these two contexts are not the same. As we discuss in Sections 2.1.5, 2.2, and 4.2.3, the stability and location of phantom images is highly dependent on a listener's location relative to the loudspeakers. To more accurately recreate a sound wave over loudspeakers that does not depend on listener location requires a different approach, one of which is called sound field synthesis. With his pioneering work in Ambisonics, Michael Gerzon was one of the leading researchers in sound field synthesis and his work remains highly important for VR audio.

2.3.1 Ambisonics: Living Up to a Past Promise

Gerzon and his colleague Peter Craven, along with members of Oxford University Tape Recording Society (OUTRS), began experimenting with surround sound recording techniques in the late 1960s. In 1971, the group made their first "soundfield" recording using four Calrec cardioid microphones arranged in a coincident tetrahedral array. They placed the microphones in the audience just above ear-level so that they could make comparisons to the sound they heard when seated in the audience. The tetrahedral microphone they developed eventually became known as an Ambisonic microphone, a technology which has become essential to VR audio and which we describe in more detail in Section 4.3.3. In their experiment, they monitored the four microphones using four loudspeakers placed around the listening area in an unusual tetrahedral arrangement as shown in Figure 2.6.

While a tetrahedral loudspeaker placement is entirely inconsistent with other multichannel loudspeaker placement standards and experiments, it was designed to work with the tetrahedral microphone. Gerzon's listening tests demonstrated that the elevated speakers affected the perception of reverberation, which helped increase the sense of realism. He also reported that even listeners who stood outside the four-channel loudspeaker array described a better sense of spaciousness in the recording than that of most other four-channel recordings (made without the tetrahedral microphone) even when standing within the loudspeaker array. One of the challenges with this recording technique, however, was getting enough channel separation. Sound sources located in front were reproduced not only from the front speakers but also to a certain extent from the rear speakers since the rear-facing microphones picked up some of the sound from the front. This reduced channel separation is also known as "overlap" (Gerzon 1971).

Clearly ahead of its time in the 1970s when Gerzon invented it, Ambisonics as an audio recording and reproduction format offers unparalleled flexibility to record and reproduce three-dimensional spatial sound over multiple loudspeakers, or binaurally over headphones. While the format never completely disappeared, standardization and technical complexity issues along with the rising popularity of cinematic formats like Dolby Surround and Pro Logic contributed to a decline in interest for Ambisonics in the 1980s. As a result, Ambisonics did not initially achieve commercial success and was largely ignored for several years except perhaps for niche audio research applications and a small number of audio recording enthusiasts. However, Gerzon's work is now more popular and widespread than ever as Ambisonics use has grown in recent years

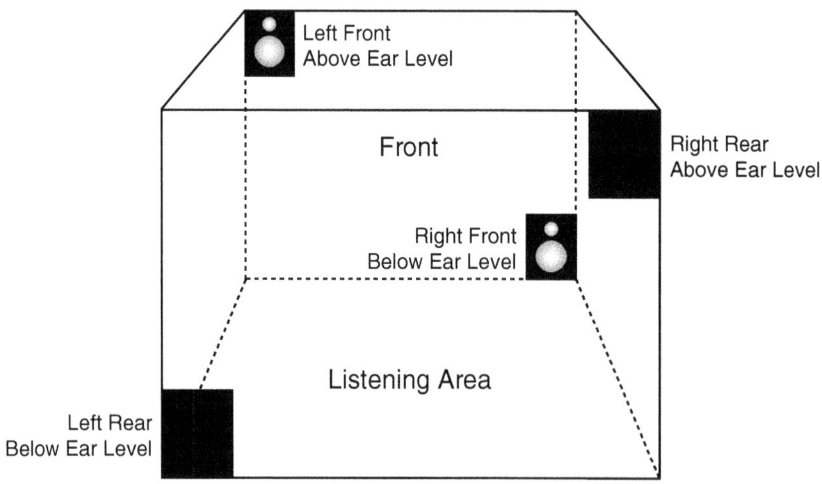

Figure 2.6 Gerzon's unusual four-channel loudspeaker arrangement, with loudspeakers positioned in corners of the room in what is known as a skew-tetrahedral layout. Adapted from Gerzon (1971).

both as a sound recording and reproduction technique and as an intermediary spatial audio format.

As we discuss in detail in Sections 4.3.3, 4.4.4, and 4.4.6, Ambisonics is a recording, encoding, and decoding format that is compatible with a variety of playback formats including VR audio. The substantial increase in 360° video, spatial audio, and VR productions, along with the commercial availability of Ambisonics hardware and software, has established the format as a primary technique for immersive media applications. As we will demonstrate in our case studies in Sections 5.5., 5.6, and 6.7, Ambisonics can be used not only for capturing immersive sound but also for rendering monophonic sounds as spatial sources in 3D space. It can also be decoded to binaural audio, so its applications are not limited to multichannel speaker arrays and can target headphone audio and VR headsets. Today, industry leaders in immersive media such as Meta and YouTube use Ambisonics as their primary format to deliver spatial audio. Gerzon was a true pioneer in the development of this technique and his legacy continues with renewed interest in Ambisonics among researchers and practitioners. In Chapters 4, 5, and 6, we will dive further into the technical implementation and applications of Ambisonics in VR audio.

2.3.2 *Wave Field Synthesis*

As an approach to sound field recording and reproduction distinct from Ambisonics, researchers at Delft University of Technology began developing wave field synthesis (WFS) in the late 1980s. Wave field synthesis is an astonishingly accurate method to reproduce an acoustic scene using many loudspeakers. Berkhout (1988) was the first to describe this audio technology which he referred to as "acoustic holography." Wave field synthesis employs a large array (usually hundreds) of loudspeakers arranged side by side around a listening space at ear level to reconstruct acoustic wave fields captured by microphone arrays. Wave field synthesis allows the placement of holographic audio images at nearly any location within or outside the loudspeaker array. Besides the accurate reconstruction of a wave field, the sweet spot is vastly expanded such that listeners at any location within the loudspeaker array have optimal listening experiences. A drawback of conventional, non-wave field synthesis, multichannel surround systems such as 5.1 or 7.1 is that sound source localization is dependent on listener location, which means that a listener needs to be in the ideal listening position (the sweet spot) to perceive sound sources at their correct locations. Listeners outside of the sweet spot in a conventional surround system localize reproduced sound sources differently than when listeners are in the sweet spot. Within a wave field synthesis loudspeaker array, on the other hand, listeners perceive a virtual sound to be originating from the same location no matter where they are located, just as with acoustic sounds in real life.

Wave field synthesis is based on a principle called Huygens-Fresnel, where, simply put, each loudspeaker reproduces some small part of the original waveform. These small waveforms (also known as spherical wavelets) that each loudspeaker produces combine to recreate a larger coherent wave. Figure 2.7 shows a diagram of a wave field synthesis loudspeaker array producing a spherical wave, which would originate from a nearby source, and a plane wave, which would originate from a distant source.

Wave field synthesis requires hundreds of closely positioned loudspeakers for accurate sound reproduction. As such, it is not only expensive but also impractical for most listening environments other than perhaps research labs or specially equipped theaters. Another

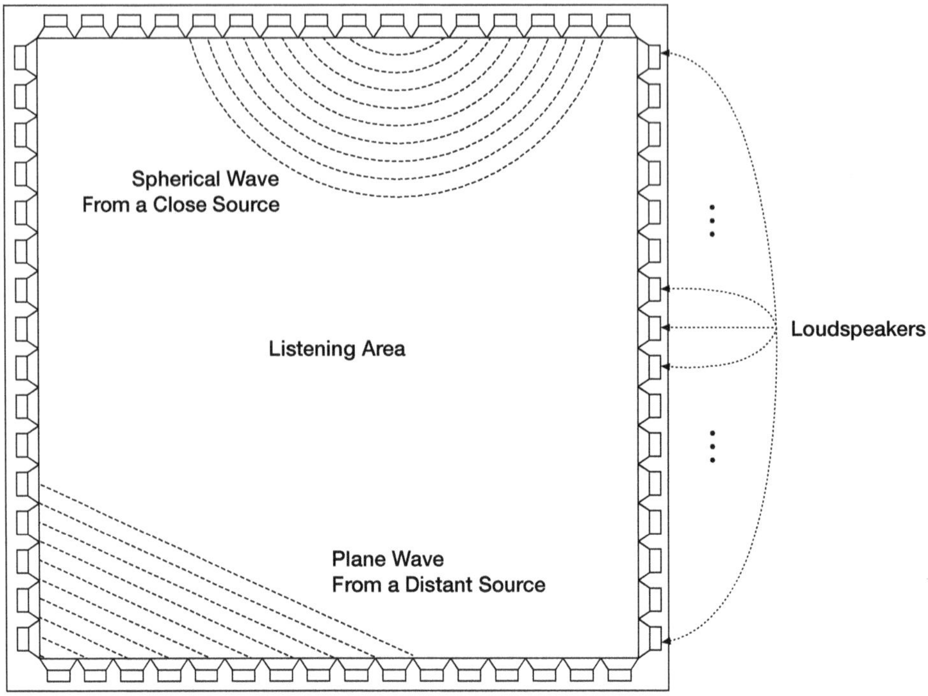

Figure 2.7 A wave field synthesis loudspeaker array showing a spherical wave from a close source and a plane wave from a distant source. A close source's perceived location is constant for all listeners and therefore its angle of arrival depends on listener location. A distant source's perceived location changes with listener location but its angle of arrival remains constant for all listener locations. Wave field synthesis can also produce virtual sound sources that appear to originate inside the loudspeaker array.

drawback is that the acoustics of the listening room may degrade the capability of a wave field synthesis system. Suitable acoustic damping may be necessary. Spatial resolution is limited due to the spacing between loudspeaker drivers. Even with loudspeakers placed side by side, the finite spacing between loudspeaker drivers limits the upper frequency of spatial resolution, also known as spatial aliasing.

Despite these challenges, the Fraunhofer Institute for Digital Media Technology (IDMT) commercialized wave field synthesis with a technology known as IOSONO. It was developed to allow up to 64 audio sources to be converted to audio objects. (We discuss audio objects in Section 4.4.2.) One practical use of IOSONO was to reproduce a 5.1- or 7.1-channel mix over a wave field synthesis system installed in a cinema. In listening tests, an IOSONO system provided more uniform panning, better image stability, and increased the size of the listening sweet spot for a 5.1-channel film soundtrack that was up-mixed to 13 channels and then mapped to a wave field synthesis loudspeaker array in a movie theater as 13 sources (Faller et al. 2013). In 2003, a 198-loudspeaker IOSONO system was installed in a 100-seat movie theater in Illmenau, Germany, and it could play back any of the existing surround audio formats including Dolby 5.1, DTS, SDDS, and stereo. IOSONO was intended for a wide range of applications including indoor and outdoor concerts, theatre productions, movie theaters, theme parks, museum exhibits, and planetariums, but other than a few installations for research projects and demonstrations, it has not been adopted as a mainstream cinema technology. Despite its remarkable spatial audio quality, wave field synthesis remains too costly and cumbersome for widespread adoption by movie theaters because of the number of loudspeakers and amount of signal processing required.

In this chapter, we provided a broad overview of spatial audio technology to illustrate its evolution from the early days of binaural sound in the late nineteenth-century Paris to more recent developments such as theatrical surround sound and sound field synthesis. Cinematic applications of surround sound have often taken priority when it comes to spatial audio research and development and that has meant a prioritization of the center loudspeaker with listeners in fixed location. Nonetheless, much of the current VR spatial audio technology draws upon the significant breakthroughs of the past. In the next chapter, we will delve into some principles of acoustics and auditory perception, especially as they relate to spatial audio perception.

Notes

1 Blumlein used the term “velocity microphone,” which is a pressure gradient or bidirectional ribbon microphone.

2 The experimental style of electronic music composition that uses recorded sounds as the raw material is known as *musique concrète* (“concrete music”).

3 The 15-track tape recorder used in the Philips Pavilion was evidently ahead of its time, compared to recording technology commercially available at the time. Recording studios at the time (1958) typically used 3-track tape recorders. It wasn’t until 1967 that Ampex mass-produced the first 16-track professional tape recorder.

4 It is interesting to note that Mercury Records uses the word “presence” in this series of orchestral recordings from the 1950s to denote the high-quality audio and realistic sound. As we discuss in Chapter 4, “presence” is an important component of VR, for which audio plays a crucial role.

Chapter 3

Acoustics and Auditory Perception

Fundamentals of Sound as an Immersive Phenomenon

A basic understanding of acoustics and psychoacoustics is essential for anyone working with immersive audio. While there are many reference texts on these subjects, this chapter focuses on their application in VR. With an understanding of sound localization, head-related transfer functions, and room simulation, sound designers can produce more realistic environments by manipulating the perceived features of a sound source or sound scene and taking advantage of the human auditory system's tendency to focus on certain aspects of a sound field while ignoring others. Furthermore, reverberation, room acoustics, and distance perception play a critical role in sound source placement within a virtual scene. In this chapter, we discuss fundamentals of sound localization and acoustic modeling to provide sound designers with knowledge that will allow them to take full advantage of the tools available to them. We first explore the basics of human auditory perception, looking at factors that govern our localization of sounds. We then focus on acoustic properties that contribute to our experience of sound in space. We end the chapter by discussing three psychological levels of hearing.

3.1 Properties of Sound

Sound is produced by vibrating objects. A vibrating object moves back and forth and causes adjacent air molecules to alternate between two conditions. In one condition, the object pushes air molecules closer together, which is compression or high pressure. In the other condition, the object pulls air molecules apart, which is rarefaction or low pressure. These alternating high- and low-pressure variations caused by a vibrating object propagate through air as a wave. Air molecules shift back and forth around their resting place but they do not travel across a space with the sound wave. Only the pressure variations move across a space. The rate at which the high- and low-pressure variations alternate determines the frequency of the sound. The unit for frequency is hertz (Hz), also known as cycles per second. One complete cycle of a given frequency has a high-pressure phase and a low-pressure phase. A high frequency sound such as 10,000 Hz goes through 10,000 high-low pressure cycles per second, whereas a low frequency sound such as 100 Hz goes through 100 cycles per second.

Sound levels are typically measured with a sound level meter which includes a microphone, an amplifier, and a meter that reads in decibels of sound pressure level (dB SPL). Loudness is the subjective perception of sound pressure level. The quietest sound humans can hear, 0 dB SPL, corresponds to the threshold of hearing. Normal speech conversation levels are around 60 dB SPL, and live concerts with amplified sound can easily reach

DOI: 10.4324/9781003048022-3

120 dB SPL. Sound levels above 120–140 dB SPL cause physical pain in the ear and can result in instantaneous hearing loss.

Sound travels about 344 meters/second in room-temperature air (20°C/68°F). The speed of sound in air is proportional to temperature, and sound travels slower in cold air and faster in hot air. A handy rule of thumb is that sound travels about 1 foot (or approximately 30 cm) per millisecond. As we'll discuss later, keeping this rule of thumb in mind can help when setting delay times in the sound design process if we know the distance we are trying to model.

Another property of sound waves is wavelength, which is the distance sound travels during one cycle of a given frequency. Wavelength varies by frequency, with lower frequency sounds having longer wavelengths. A sound's wavelength becomes relevant as we consider objects that a sound might encounter in a virtual space since an object's size relative to the sound's wavelength determines whether a sound bounces off the object, bends around it, or passes through it. To find the wavelength of a sound at a particular frequency, we can simply divide the speed of sound (344 meters/second) by this frequency:

$$Wavelength\,(m) = \frac{Speed\ of\ Sound\left(\frac{m}{s}\right)}{frequency\,(Hz)}$$

The audible frequency range for humans is from approximately 20 Hz to 20,000 Hz for those with excellent hearing such as young children. Sensitivity to high frequencies declines with age and due to exposure to loud sounds. There is a wide variation in wavelength from the lowest audible frequency to the highest. Based on the formula for wavelength, we can compare the wavelengths of the lowest and highest audible frequencies:

Frequency	**Wavelength**
20 Hz (lowest)	17.2 m
20,000 Hz (highest)	0.0172 m

In other words, sound travels 17.2 m (677 in.) to go through one cycle at 20 Hz, and only 1.72 cm (0.68 in.) to go through one cycle at 20,000 Hz. Inversely related to frequency is period (T), which is the amount of time it takes to go through one cycle of a given frequency (f), and can be calculated with the following formula:

$$T\,(seconds) = \frac{1}{f\,(Hz)}$$

3.2 Basics of Human Sound Localization

The human auditory system relies on three main binaural cues to localize sound: interaural level differences (ILD), interaural time differences (ITD), and direction-dependent filtering by the outer ears, head, and torso. Through many listening experiments, researchers have found that humans rely mainly on interaural level difference cues to localize high frequencies (above about 1,500 Hz) and ITD cues to localize low frequencies (below about 1,500 Hz). The auditory system uses different cues for low- and high-frequency localization because sound wavelengths vary with frequency yet the distance between our ears is fixed. If we assume the diameter of a human head is 8.75 cm (3.4 in.), the maximum interaural time difference is estimated to be 763 microseconds (0.763 milliseconds) when a source is at 90° to one side in a spherical model of a human head (Hartmann 2021). Lord Rayleigh was the first to analyze the role of interaural level differences in localizing high frequencies and interaural time differences in localizing low frequencies. His finding of two different sound localization cues became known as the "duplex theory" of sound localization (Rayleigh 1907).

3.2.1 Interaural Level Differences

Since low-frequency sounds have longer wavelengths, they bend or diffract around the head. Therefore, no matter the direction low frequencies may originate, they are likely to reach both ears at about the same level. This typically results in only minimal level difference between the ears for low frequencies, which means interaural level difference is an ineffective cue for low frequency localization.

High-frequency sounds, on the other hand, have short wavelengths relative to the size of the human head, so they do not bend around the head. Instead, high frequencies are blocked and reflected by the head. The result is that high frequencies arriving from one side create an acoustic shadow over the more distant (or contralateral) ear. An acoustic shadow means that the level of sound reaching the more distant ear can be significantly attenuated relative to the closer (or ipsilateral) ear for high frequencies only. Spherical head models show, for example, that 4,000 Hz can have nearly 16 dB of interaural level difference (Hartmann 2021).

3.2.2 Interaural Time Differences

Another localization cue we rely on is interaural time difference, that is, the difference in the time for a sound to reach one ear versus the other. While we utilize interaural time difference cues for low frequencies (below about 1,500 Hz), these cues become ambiguous at high frequencies. High frequencies above about 1,500 Hz have wavelengths that are shorter than the distance sound travels to reach one ear compared to the distance traveled to reach the other ear. As a result, the time it takes for sound to travel from one ear to the other represents more than one cycle of phase delay with high frequencies. If phase delay is more than one cycle, interaural time difference becomes an ambiguous cue. The auditory system does not know if only one cycle or a hundred cycles have passed when the delay time is more than one cycle. Therefore, interaural time differences are an ineffective cue for high frequencies.

We can mimic the interaural time difference effect on localization by playing a transient signal, such as a click, to the left and right ears over headphones. With no time or amplitude difference between the two channels, listeners localize the click inside the head, an effect also known as in-head localization or lateralization. If we delay the left channel click, for example, by some amount under 1 millisecond (ms), listeners localize the click towards the right (the leading side if the left channel is delayed). Recall how models estimate that the maximum interaural time difference is less than 0.8 ms. As we increase the delay time up to about 1 ms, listeners perceive the sound source to move towards the side that has not been delayed. At timing differences below 1 millisecond, listeners do not perceive the clicks at the two ears as separate events. Over headphones, the clicks coming from left and right are perceptually fused when the time difference between them is less than about 2 ms. Listeners perceive that there is only one click, and it is located to the right when the left channel is delayed (or vice versa when the right channel is delayed). This phenomenon is known as the law of the first wavefront or the precedence effect as described by Wallach, Newman, and Rosenzweig (1949). As the delay time increases from about 2 ms to about 5 ms the sound image begins to pull apart and listeners begin to hear the delayed channel as a widening of the sound image. A listener's ability to perceive the two as a separate event starts when the time of arrival difference between a click and a delayed version of the click is more than about 5 ms. Importantly, the minimum time delay in which listeners can perceive two separate sound events—the leading sound and the lagging sound—might be longer than 5 ms and it depends on the characteristics of the sound source itself. Listeners perceive two separate events (an echo) with a shorter time delay when the sound source is a click. If the sound source is more steady-state or sustained such as a held note on a violin, the time delay between the direct (leading) sound and the echo (lagging sound) needs to be much larger to be perceived as an echo because of masking. With a sustained sound, the leading (direct, non-delayed) sound covers up or masks the lagging (indirect, delayed) sound, making it difficult to hear a distinct echo unless the delay time is significantly longer.

3.2.3 *Filtering by the Outer Ears*

The third sound localization cue, after interaural level and time differences, uses acoustic filtering to help determine a sound source's elevation. When sound arrives at a listener, it bounces off the small folds of the outer ears (or pinnae), the head, and the shoulders before it reaches the ear drums. This series of tight reflections causes the sound to be filtered acoustically, which allows a listener to judge a sound source's elevation. Sounds at different elevations will undergo different filtering based on the way sound bounces off the folds of the outer ear. This filtering also helps externalize sound, which means that listeners perceive acoustic sounds as originating from outside the head. Sound externalization is the perception that a sound source is outside the head, which is the normal state of auditory perception for natural, acoustic listening when we are not listening over headphones. Although the fact that listeners perceive external sounds outside the head may seem obvious, unless sound bounces off the outer ears, head, and torso, listeners tend to localize it inside the head if both ears receive identical signals, such as through headphones, which, as we mentioned above, is called in-head localization or lateralization.

Since headphones are the primary sound reproduction medium for VR, in-head localization is a possible artifact of the experience. We want to be aware of this possibility because users who experience in-head localization may feel a diminished sense of presence and immersion in VR. Binaural signals are generally externalized when they include the auditory cues that are normally present in natural listening and when they fit within a listener's experiences (Best et al. 2020). Klein et al. (2019) report that external localization in binaural playback is mandatory to create a plausible auditory illusion, but that switching between virtual rooms or switching from a real room to a virtual one can create in-head localization until listeners have time to adapt to the new room acoustics. The effect is known as the divergence effect, which can be diminished with ongoing exposure to the room acoustics condition.

3.2.4 Sound Localization Issues

The three main sound localization cues—interaural level differences, interaural time differences, and filtering by the outer ears—do, however, have a shortcoming that can cause localization errors. A sound source's perceived location can be ambiguous at certain locations because some locations share interaural time differences and interaural level differences. These locations happen to occur along a cone shape that extends out from the middle of the head to each side. The term "cone of confusion" is used to describe these ambiguous locations. Sounds located along the cone of confusion are equidistant from the left and right ears and thus produce the same interaural time and level differences for a listener. We call it the cone of confusion because it causes listeners to perceive a sound in front as arriving from the rear or vice versa, also known as "front-back confusion." It also causes listeners to perceive a sound above as arriving from below or vice versa, which we call "up-down confusion." Figure 3.1 illustrates the cone of confusion.

In natural listening, it is usually easy to resolve these kinds of sound source location ambiguities by turning our heads slightly. Small head turns can change the interaural time differences enough for a listener to determine a source's location. When we hear a sound to the side or behind us, our natural tendency is to turn our head towards the perceived direction of the sound source to confirm the location with our eyes. The resulting change in head rotation affects the interaural differences. Turning towards the location of the sound will decrease the interaural time and level differences until the source is directly front of us and interaural differences are minimized.

Our perception of sound source azimuth (angular location) is normally more robust than our perception of elevation, mainly because of the localization cues involved. When a sound source moves from left to right, interaural time and level differences change, but when it changes elevation without changing azimuth, interaural time and level differences remain constant because the distance to each ear does not change. The outer ears or

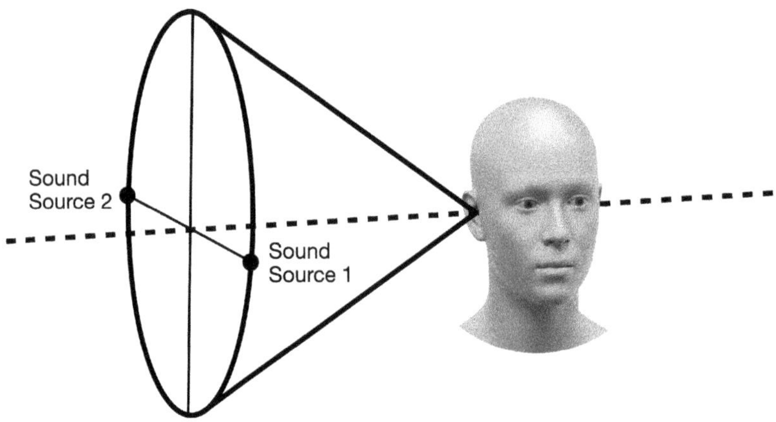

Figure 3.1 A diagram showing the cone of confusion, which is an imaginary cone extending out from the middle of the head. Sound sources located any place along the surface of the cone are equidistant from each ear and have identical interaural time and level differences. The cone of confusion causes listeners to perceive a sound in front to be located behind or vice-versa (called "front-back confusion"), or to perceive a sound above as arriving from below or vice versa (called "up-down confusion"). Adapted from Fischer, Caversaccio, and Wimmer (2020).

pinnae provide cues for elevation because of the way they filter sound and change its timbre, but these filtering cues are not as strong as interaural time and level differences. Furthermore, pinnae filtering cues for elevation vary from person to person because no two sets of outer ears are identical. Thus, elevation control is more difficult in spatial audio. Distance control typically requires the inclusion of room simulation, such as reflections and reverberation, as well as control of a sound source's amplitude. By adding a time delay between direct and indirect sound, we can provide listeners with a sense of distance and space. We discuss reflections and reverberation and their impact on distance perception further in Sections 3.3 and 3.5.

3.3 Direct and Indirect Sound

When sound travels from its source to a listener, it takes one of two different types of paths: direct and indirect. Direct sound takes the path straight from the source to the listener, which assumes there is a clear line of sight between the source and the listener (i.e., there is no object blocking the path between the two). Sound from the same source that arrives from any other direction is indirect sound, since it has been reflected off one or more surfaces or objects before reaching the listener. Direct sound reaches the receiver first since the direct path is shorter than any route that reflected sound may take. Arriving slightly after the direct sound, each reflection is essentially a delayed, if not also slightly filtered, version of the direct sound.

We usually divide indirect sound into two categories: early reflections and reverberation. Early reflections arrive just after the direct sound, have typically bounced off one or only a small number of surfaces, and are more separated from each other in time than reflections are to each other in the reverberation tail. Nearby hard, flat surfaces such as the ceiling and side walls of a space can produce strong early reflections. Figure 3.2 is a simplified impulse response showing the time of arrival of direct sound, followed by early reflections, and reverberation decay as we might measure it in an acoustic space or artificial reverberation. The direct sound in an impulse response is a short click and is shown in the figure by the first vertical line in time. Early reflections are shorter vertical lines to show that they are delayed versions of the direct sound but at lower amplitudes. The reverberation component of the impulse response diagram below is simplified as a solid triangular block showing that the sound level gets quieter over time as the reverberation decays. Reverberation is a dense collection of reflections that begin to overlap with one another. There is no precise dividing line between early reflections and reverberation, but in general we consider the transition from early reflections to reverberation starts when individual reflections are no longer distinguishable. Usually, this occurs within about the first 100 to 150 milliseconds after the direct sound.

The human auditory system is sensitive to the direction of arrival of early reflections, with side reflections contributing to the perception of width and spatial impression (Barron and Marshall 1981). Individual reflections are usually quieter than the direct sound since reflections travel further than the direct sound, and the further sound travels the more its level decreases. The frequency content of reflections is also affected by the surface material from which the sound reflects. For example, surfaces like carpet, fabric upholstery, and heavy drapes absorb high frequencies more than low frequencies, which means reflections in a carpeted room will likely have reduced high frequency content than reflections from a room with a hard floor. Reflections usually begin to arrive within 20 ms of the direct

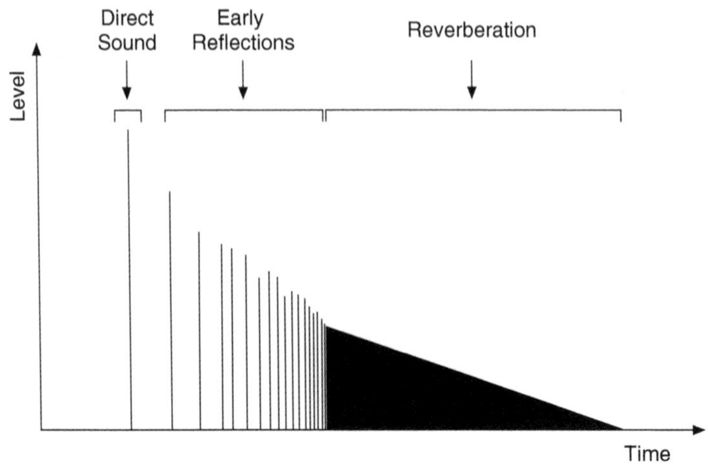

Figure 3.2 A simplified diagram of an impulse response showing the decay of sound, from early reflections through reverberant tail, in an acoustic space or artificial reverberation processor. The x-axis is time and the y-axis is level. Early reflections, which occur after the direct sound but before the reverberation, are shown as discrete (or separate) copies of the original direct sound but delayed in time and at a lower level. The reverberation is shown on the right side as solid black to represent the much higher density of reflections at that point in time. As we move further in time away from the original direct sound, reflections get closer and closer together as they decay in level.

sound. Since the direct and reflected sounds are so close in time, they are grouped together and perceived as one sound. However, this grouping depends on the nature of the sound, as we mentioned earlier. Two short clicks can be heard as separate events if they are spaced apart by only 5 ms, but a more complex sound such as speech or music can be separated by up to 40 ms before we hear the direct and reflected sound as separate events. If the direct sound and reflections are grouped together, their location is determined mostly by the location of the first arriving sound, thus the precedence effect is relevant to localization in real acoustic spaces or when we simulate the acoustics of a space, and early reflections are present. In the time following a sound's onset, sound continues to bounce off nearby surfaces and, assuming the room has sufficient volume and reflective surfaces, early reflections transition into audible reverberation, as reflections overlap more and more in time. As we move later into the reverberation tail, reflected sound becomes more diffuse, meaning that the reverberant sound arrives from all directions equally. In the reverberation tail, statistical properties like reflection density and the rate at which sound decays become more important than the timing of individual reflections since individual reflections are indistinguishable in the reverberation tail (Hacihabiboglu et al. 2017). The sound level in the reverberation tail decays naturally as the reflected sound energy drops over time. We define the reverberation time (or RT60), as the amount of time it takes for sound level to drop by 60 dB after the direct sound has stopped.

If we change location and increase our distance from a sound source, the direct sound level will be lower (since the sound must now travel further to reach us), while the indirect sound (consisting of reflections and reverberation arriving from all directions) level may remain relatively constant, depending on the room. Direct sound level decreases as it travels away from its source at a rate of about 6 dB per doubling of distance, also known as the inverse square law. At some distance away from a source in a room, the direct and indirect sound are equal in level. This location, where the direct and indirect sound levels are equal, is known as the critical distance or reverberation radius. The critical distance depends on the size of the room and the reflectivity of the surfaces. In a large room with highly reflective surfaces, the critical distance may be relatively short compared to a room with highly absorptive surfaces. Speech intelligibility can suffer greatly as we move past the critical distance since reflections and reverberation mixed with speech impact intelligibility, which is something to keep in mind for VR sound design and the use of physical models for acoustics simulation. Directional loudspeakers and loudspeaker fills are often used for speech reinforcement in live spaces to help with intelligibility by augmenting the level of the direct sound. Further away from a source as we move past the critical distance, the direct sound level drops below the indirect sound level. While it is often impractical to calculate the critical distance of the room in which we are recording, it is a useful concept to keep in mind when placing microphones around sources for recording and when creating virtual spaces.

Direct and indirect sound levels, as well as the critical distance, are relevant to our perception of a sound source's distance from us. In the real world, for those without visual impairments, we may be able to see how far away a source is, assuming it's in a line of sight. When creating virtual spaces, the sound design and room acoustics model should be consistent with the visual layout so that users perceive sound source location unambiguously.

In addition to reflections and reverberation, room modes can be prominent in smaller acoustic spaces. Also called standing waves, room modes are acoustic resonances that accentuate certain frequencies, mainly in the low frequency range. Frequencies get accentuated when the distance between parallel walls or between the ceiling and floor is some multiple of a frequency's wavelength. When half the wavelength equals the room dimension, the reflected wave is in phase with the incoming wave, and it reinforces itself. Frequencies whose wavelengths are not a multiple of the room dimensions do not get reinforced as room modes. Practically this means that some frequencies get amplified by the room but only at certain locations within the room. Standing wave frequencies are not evenly spread throughout a room. At certain locations in a room, at what are called the nulls of the room mode, the standing wave frequency may be nearly completely attenuated. At other locations, it might be amplified by 6 dB because it is constructively interfering with itself and doubling in amplitude. Resonances occur not only in a single dimension between parallel walls but also in two and three dimensions.

3.4 Binaural Audio and Head-Related Transfer Functions (HRTFs)

Despite the numerous surround sound and multichannel loudspeaker array configurations that have been devised, only two audio signals—one for each ear—are needed to create a three-dimensional auditory perspective. We are immersed in sound every day and in natural hearing we localize acoustic sounds around us using just two ears. After the ears receive sounds, the brain processes the auditory signals transmitted from the inner ears. Because ears are positioned on opposite sides of the head and therefore not situated in the same physical location, the sounds received by the two ears differ. As we discussed above, these interaural differences and outer-ear filtering help us localize sound. The transformation that sound undergoes through the interaural time and level differences and outer-ear filtering that are introduced when sound travels from a source to our two ear drums is called a head-related transfer function (HRTF). Said another way, a head-related transfer function is the frequency response of a person's ears for a given sound source location, and it includes inherent interaural time and level differences and outer-ear filtering. The outer ears, ear canals, head, and torso all contribute to the filtering of sounds before they reach the ear drum due to the way sound reflects off or diffracts around these surfaces. Head-related transfer functions vary according to sound source location, as well as by the shape of the outer ears and head dimensions, which vary from person to person. Each sound source location has a unique head-related transfer function, each ear will have its own head-related transfer function, and head-related transfer functions vary from individual to individual because of differences in outer ear shapes and differences in dimensions of the external ears, head, and torso. Pioneering research on sound localization in binaural audio by Morimoto and Ando (1980) and Wightman and Kistler (1989) found that if we replicate the sound that would have arrived at the ear drums for a given sound source location (which we can do by playing these signals over headphones), listeners localize sound sources at their virtual locations with varying degrees of success.

3.4.1 *Deriving HRTFs*

Quantifying the transformation that an audio signal undergoes from sound source to ears in the form of a frequency response curve means that we can apply head-related transfer functions to audio signals like we do with a filter. When we filter audio using a head-related transfer function, localization cues (interaural level differences, interaural time differences, and spectral filtering) are applied to the audio for a given sound source location. If we measure our own head-related transfer functions and listen back over headphones to audio processed with them, we localize the sound source at the location in space from which the measurement was taken. What is incredibly useful about head-related transfer functions, especially for VR, is that we can filter any sound with head-related transfer functions to control the virtual location of a sound source over headphones. Head-related transfer functions are an essential component of VR audio because they allow sound to be spatialized over headphones as binaural audio. We can think of head-related transfer functions like acoustic fingerprints because although there are a lot of similarities among them from different people, everyone has their own unique HRTF. Head-related transfer functions vary because the shapes and dimensions of the outer ears (or pinnae), head, shoulders, and torso are unique for each person. Although the general shape and features of the pinnae are common to almost everyone, head size and outer-ear dimensions vary from person to person. These small physical differences can result in HRTFs that create sound localization errors, produce tonal coloration and high frequency harshness, and reduce sound externalization. A person's HRTF represents the connection between their specific anatomy—shape and dimensions of the outer ears and head—and their perception of sound source location in space. Not only are HRTFs unique to each person, but HRTFs also vary with sound source location.

We use HRTF filtering for VR audio to provide 3D audio over headphones that mimics or recreates sound localization cues of natural hearing as closely as possible. Ideally the aural experience of VR audio should be indistinguishable from the experience of hearing natural sounds in everyday life. Without using our own HRTFs, the experience can be less than optimal. When we use generic HRTFs, such as those from a binaural head, the accuracy of localization and sense of realism varies from person to person, depending mainly on how closely the generic HRTF matches an individual's HRTF. Even in natural, acoustic hearing, sound localization is not always accurate due to the cone of confusion, so we often use slight head movements to help localize sounds and resolve front-back ambiguities. It is for this reason and because we are often working with non-individualized HRTFs that head tracking is important for VR applications, a topic we discuss more in Section 4.2.1. Work by Gupta, Barreto, and Ordonez (2002) indicates that individuals with particularly protruding ears experience less localization errors on the cone of confusion. HRTFs from protruding ears, when applied to other listeners, can improve front-back differentiation.

To experience spatialized audio with our individual HRTFs, we must have our HRTFs measured acoustically. HRTF measurement is usually conducted with the listener seated and their head fixed in one position inside a specialized sound laboratory called an anechoic chamber. Miniature microphones are placed in their ear canals to record the incoming sound produced by either an array of loudspeakers or a single, movable loudspeaker that generates sound from many different locations, one location at a time, in the 360° spherical space around the listener. The ear microphones record audio from each location, and then the recorded audio from each source location is mathematically converted to represent the difference between the original audio fed to the loudspeaker and the audio recorded by the in-ear microphones. The difference between these two signals, the original source audio and the recorded in-ear audio, gives us the HRTF.

3.4.2 Generic HRTFs

To support research in sound localization and 3D audio, audio scientists Bill Gardner and Keith Martin at the MIT Media Lab have made generic sets of HRTFs measured using dummy heads publicly available. They sampled 710 positions around a KEMAR (Knowles Electronics Mannequin for Acoustic Research) dummy head in an anechoic chamber at elevations from -40 degrees to +90 degrees (Gardner and Martin 1995). The KEMAR HRTF dataset from MIT has become one of the most widely used HRTF measurements for research in sound localization and 3D audio. In 2001, researchers at University of California, Davis, Center for Image Processing and Integrated Computing (CIPIC) Interface Laboratory measured the HRTFs of KEMAR along with 43 human subjects and made the dataset publicly available to help understand person-to-person differences in HRTFs (Algazi et al. 2001). In 2003, IRCAM in Paris made their Listen HRTF database publicly available (Warusfel 2003). More recently, Meta Reality Labs Research released a high-resolution database of 78 HRTFs that is also freely available to download (Warnecke et al. 2024).

While these HRTF databases are highly useful, there is one main drawback when we use HRTF measurements from someone else or a dummy head such as the KEMAR. Listening to sounds filtered through someone else's HRTF or through a generic HRTF will often affect localization ability and how well sound is externalized. As we mentioned earlier in the chapter, when sound is not externalized, listeners perceive it as originating inside the head, an effect known as lateralization. As head-related transfer functions vary from person to person, many studies have shown that when listening to a recording processed through HRTFs that are not our own, that is, non-individualized HRTFs, our ability to localize and externalize sound is compromised at least for static listening conditions. If a listener happens to have measurements of their own HRTFs and they process sound through their personal HRTFs, the result is like natural hearing. Results of listening tests conducted by Jenny and Reuter (2020) indicate that individualized HRTFs sound more realistic, allow easier sound localization and externalization, and have a more natural timbre than with non-individualized HRTFs.

3.4.3 *Listening to Generic HRTFs*

Listening to the HRTFs of another person is like listening through their ears, but people differ in head dimensions and the specific shape of the outer ears. If we listen to audio processed with someone else's HRTFs, the specific interaural timing, amplitude, and spectral cues will likely differ from our own. We develop localization capabilities over time using our own ears, head, and torso, that is, our own HRTFs. We have lived with the inherent interaural differences and pinna filtering of our own ears our whole lives, so when any aspect of these features changes, our localization accuracy may change too. If we want listeners to have a truly immersive VR experience in which they forget that they are listening to sound reproduced over headphones, we need to match the HRTFs used in the binaural playback system as closely as possible with the listeners' individual HRTFs. If that is not possible, which it's not in most cases, we can use some additional signal processing to mitigate the effect.

To illustrate this point, it is worth listening to the web-based immersive audio design tool Inviso, a screenshot of which is shown in Figure 3.3. This platform, like most commercially available spatial audio tools, uses a generic HRTF to spatialize sounds in a virtual 3D environment for listening over headphones. You can find the Inviso scene depicted in Figure 3.3 in this book's project repository hosted at vraudio.org under Chapter 3. In this scene, there are two automated moving point sources: one that circles around the user in the horizontal plane and one that moves up and down in front of the user. Play the first sound to see how well you can localize it when it's all the way to the left or right and when it's right in front of you or behind you. Then pause this sound and play the second one to see if you can hear a difference as the sound moves on the vertical axis. When you listen to the Inviso scene, is the sound source located where you expect it to be as indicated by the visual display? Do you notice any ambiguity in the sound source's location compared to its location in the visual display? You may experience front-back or up-down confusions or localization errors because your individual HRTFs may not precisely match those that are averaged in the generic HRTFs.

3.4.4 *Approximating Individualized HRTFs*

HRTF measurement is a time-consuming process, and it is obviously not feasible for everyone who wants to experience VR audio to go through this process in an anechoic chamber to take individualized HRTF measurements. Because of the difficulty and expense of HRTF measurement, researchers have been working on other ways to

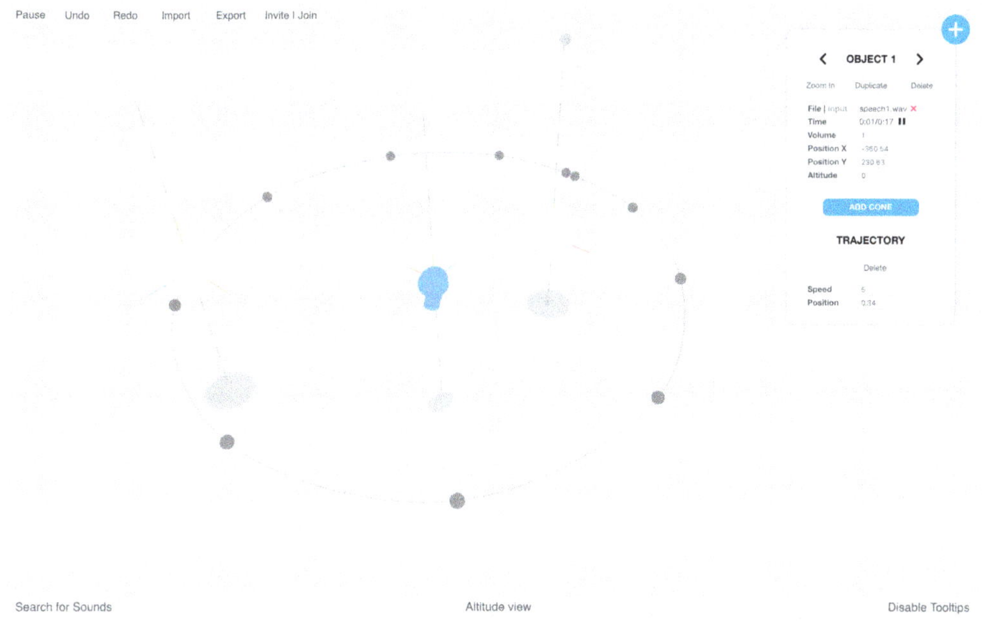

Figure 3.3 A virtual auditory scene consisting of two sounds created in the web-based immersive audio design tool Inviso (accessed at inviso.cc). One sound source in the scene circles around the user and the other one moves up and down. The head model (in blue) in the middle represents the vantage point from which we hear this scene. The HRTFs used to spatialize the sound are generic so listeners may experience front-back or up-down confusions or other errors in sound source localization. You can find this example in Inviso_BinauralText.zip, which can be found at vraudio.org under Chapter 3.

generate individualized HRTFs or to determine how generic HRTFs might be modified to optimize localization and spatialization for most users. One alternative to individual HRTF measurement uses still images or videos of users' heads and ears to either generate individualized HRTFs or to select the best match from an existing set of HRTF data sets. While computer vision can be used to extract the depth images needed for this process, the depth cameras that are now used in commercial smart phones are also being used for it. As a result, some companies invite users to record and upload images and video of their own ears and head for HRTF measurement. For example, Meta has developed a method to generate individualized HRTFs based on photos or videos that users upload to a server, as reported in one of their patents (Brimijoin et al. 2020). Based on photographs of an individual's anthropomorphic features—specifically, the shape and dimensions of the outer ears and head—the Meta system searches a database for an HRTF template that matches the individual's HRTF as close as possible and then modifies the template to create an approximation of the individual's HRTF. Sony initially introduced their 360 Reality Audio technology in 2019 with generic HRTFs but now offers users the ability to personalize their HRTFs. With the Sony Sound Connect smartphone app users take pictures of each ear, and the app analyzes the ear images and optimizes the spatial audio experience based on the shapes of a user's ears. Researchers at Microsoft are working on a method to measure a listener's HRTFs using headphones. The technique uses a microphone attached to the exterior of one headphone to measure sound propagation around the head, a magnetometer to measure magnetic field intensity attenuation by distance, and an accelerometer to help estimate the distance between the ears (Islam and Tashev 2020). Dolby announced in 2022 that they had begun beta-testing their Personalized HRTF measurement application. Using an iPhone that supports Apple Face ID or an Android phone with Android Pi (9) or later, a user takes a short 360° video of their head (including outer ears) and upper torso and uploads it to a Dolby server where it is converted to a personalized HRTF (PHRTF). The goal is to improve the binaural rendering experience (Dolby Laboratories, Inc. 2022; *Sound on Sound* 2022). In 2022, Apple began offering users the ability to listen with what they call "Personalized Spatial Audio." Using a compatible iPhone with a depth camera, a user can take an image scan of their face and ears to create the option for Personalized Spatial Audio that can be used with certain models of Apple headphones. According to an Apple patent, it appears that the technology records a 3D image of a user's head and then selects the most appropriate HRTF from a database of pre-measured HRTFs (Satongar et al. 2020). Meta has also worked to improve upon generic HRTF models available and introduced their Universal HRTF for the Meta Quest in 2023. They captured a data set of nearly 150 high-quality HRTFs and created their own generic model from this data set. They report that based on their listening tests, their Universal HRTF offers improved localization and more natural timbre with less coloration and filtering (Meta 2023).

3.4.5 HRTFs and Sound Source Distance

Sound source proximity can also be a factor affecting sound localization in binaural audio since, as Brungart (2002) describes, studies have shown that when sound sources are in the near field, that is, less than 1 m (3 ft.) away, HRTFs change substantially with distance. The most significant changes to near-field HRTFs are in low frequency interaural level differences (ILD). As we discussed earlier in this chapter, we generally do not rely on ILDs for low frequency localization most of the time since sound sources are usually further than 1 m away. For closely positioned sources, however, the inverse square law becomes relevant and has implications for ILDs. The inverse square law states that for every doubling of distance, sound level drops by 6 dB in a location with no reflective surfaces present, a condition known as a "free field." For example, if a sound source is 70 dBSPL when it is at 2m away, it will be 64 dBSPL when it is 4 m away, and 58 dBSPL when it is 8 m away. The inverse square law assumes an omnidirectional point source, which radiates sound equally in all directions, and that the source is in a free field, an environment with no reflective surfaces nearby. Since sound expands out spherically in all directions from a source, and the energy of a wave gets spread over an increasingly larger surface area of an expanding sphere, the sound level drops proportionally. Sound intensity (I) drops proportionally to the inverse of the square of the distance (r), also known as the inverse square law, represented by this formula:

$$I = \propto \frac{1}{r^2}$$

A more practical way to calculate the drop in level due to distance is with the following formula that compares distances:

$$Lp_{r2}(dB) = Lp_{r1}(dB) - 20 \cdot \log_{10} \frac{r_2}{r_1}$$

where, Lp_{r1} = the known sound pressure level (in dB) at the first location
Lp_{r2} = the unknown sound pressure level (in dB) at the second location
r_1 = distance from the sound source to the location of the known sound pressure level
r_2 = distance from the sound source to the location of the unknown sound pressure level

Distances can be in any unit as long as both r_1 and r_2 are in the same unit.

For sources close to the head, if the distance from the sound source to the opposite (more distant or contralateral) ear is more than double the distance to that of the closer (or ipsilateral) ear, it could result in more than 6 dB of ILD for all frequencies, including low frequencies. Backing their belief that near-field HRTFs are important for optimal spatial audio reproduction, Meta made near-field HRTFs available in the Rift SDK in 2017 to enhance the VR experience and make it more realistic (Oculus VR 2017).

3.5 Immersion and Acoustic Modeling

An ideal VR experience provides the user with a sense of immersion and envelopment, promotes the suspension of disbelief, and matches users' expectations in a natural way through the presentation of appropriate auditory and visual cues. Auditory cues include sound source locations or panning, the presence and timing of early reflections and reverberation, and filtering. Visual cues include the shape and dimensions of the space or environment and the placement of the sound-producing objects. To evaluate audiovisual consistency, we can examine whether a virtual environment sounds the way it looks. Do the sound source locations in the audio and visual domains match? Does the size of the space visually match the sound of the reflections and reverberation? Congruence between audio and visual cues is crucial for maintaining a user's sense that they are immersed in a virtual world (Jot et al. 2021).

We can predict the propagation of sound through a space with acoustical models such as image-source or ray tracing, and these models can be used to generate or render reflection patterns and reverberation in a virtual environment. One challenge with room simulation for VR is the dynamic nature of the medium in which listeners are free to move relative to sound sources, which themselves may move. In a real acoustic space, sound reflection patterns and timbre change as sound sources and listeners move within that space. To maximize the realism of a sound scene in VR, the acoustic simulation, including reflections and reverberation, must update dynamically as listeners and sources move within the virtual environment.

3.5.1 *Room Acoustic Modeling Techniques*

Sound propagation models often utilize a geometric room acoustic modeling technique such as an image-source model, ray tracing, or a hybrid of the two. An image-source model is a relatively efficient and easy way to simulate early reflections in a rectangular shaped room, as described in a classic paper by Allen and Berkley (1979). In an image-source model, simulated reflections are *specular*, meaning that sound is reflected from a room boundary (i.e., a wall) like a ray of light is reflected from a mirror; the angle of incidence equals the angle of reflection therefore the sound is not scattered. Distinct from specular reflections, diffuse reflections are those that are reflected or scattered in multiple directions simultaneously, rather than in just one direction. The limitation in modeling only specular reflections may be a shortcoming since diffusive surfaces are common in real life—not every surface is hard and flat—and simulating diffusion in a virtual environment may help with realism.

In an image-source model, the number of room boundaries a sound hits before arriving at a receiver determines the "order" of the reflection. For example, a third-order reflection hits three surfaces before reaching the listener or microphone, whereas a first-order reflection hits only one surface. Figure 3.3 shows the path of the direct sound and three example reflection paths in a two-dimensional image source model in which there is one sound source and one receiver or listener. The rectangle represents a room with a sound source (denoted by "X") and a receiver (denoted by "O"). Virtual locations of the image source reflections are denoted by "X." As the image source model reflection order increases to higher orders, computational intensity grows significantly. Therefore, this method is most appropriate for generating lower order reflections to simulate early reflection patterns, but it is less appropriate for generating a reverberation tail. It is also not the best method for generating reverberation tails since there are more efficient methods to do that as we discuss below.

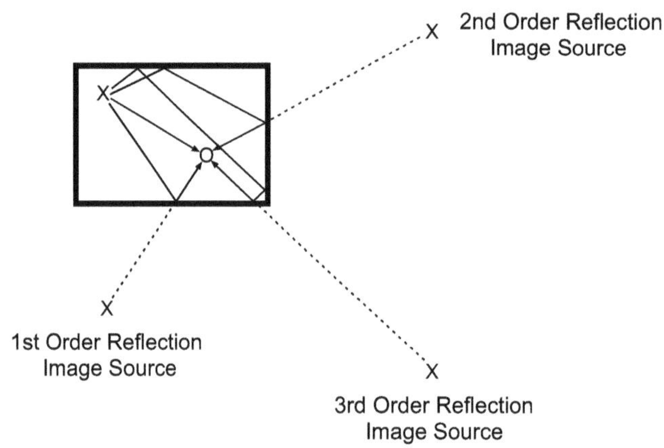

Figure 3.4 A diagram of a two-dimensional image-source model showing the spatial arrangement of the image space. The bold rectangle represents the top view of a rectangular room, with the source shown as "X," and the receiver (i.e., listener or microphone) shown as "O". Image sources of three reflections are represented each by an "X" at their respective virtual locations. The solid line with an arrow going straight from the X to the O in the bold rectangle shows the direct sound path from source (X) to receiver (O) within the room. Three example reflection paths are indicated by solid lines that travel from source to the walls and then to the receiver. A solid reflection line when unfolded extends out to its image source's respective virtual location. The dotted lines show examples of first-, second-, and third-order reflections from the virtual image sources.

Ray tracing is another common method to model the propagation of sound in a room. Ray tracing treats sound like rays of light that spread out into a room in all directions from a sound source. Each ray travels from a source and is subsequently reflected, absorbed, refracted, or diffracted each time it hits a surface. Sound rays' interactions with each surface alter the reflected sound's characteristics (such as level and frequency content), just like in a real acoustic space. Sound rays continue to reflect off surfaces until a defined end condition is met, such as a set number of reflections or the sound reaching a listener. Ray tracing can produce a realistic sounding model of sound in a room, but it loses temporal resolution for late reflections because only a finite number of rays can be traced. In other words, late reflections become increasingly spread out in time as we move further in time beyond the initial, direct sound.

Geometrical models such as image-source and ray tracing incorporate sound attenuation due to distance traveled according to the inverse square law, high frequency attenuation due to air absorption (typically simulated with a high-shelf cut or low-pass filter), and filtering due to frequency-dependent absorption by each surface from which the sound is reflected before it arrives at the receiver. As sound propagates through air, high frequencies are typically absorbed more than mid or low frequencies, thus we can model this effect by filtering high frequencies. Surfaces do not typically reflect all frequencies equally so we can filter reflected sound to approximate this effect. When sound hits a surface and reflects from it, some frequencies may be attenuated more than others.

When an object is between a sound source and a listener, and therefore blocking the sound's direct path, the sound is blocked or occluded. Typically, higher frequencies are attenuated more than low frequencies since low frequencies will bend around the object. This means that we can usually still hear the sound source, but it may be muffled because of the loss of high frequencies.

Meta announced in 2024 that they now include Acoustic Ray Tracing capabilities for Unity and Unreal as part of the Audio SDK (Meta 2024) to simulate the way sound propagates based on the actual geometry of the virtual environment. It adapts dynamically in real time to movement of the users and objects to create a realistic acoustic room simulation, and models sound diffraction or bending around objects, sound getting blocked by objects (also known as occlusion), and the changes that sound undergoes when it passes through different materials. We discuss room effects and occlusion further in Sections 6.5.5, 6.6.2, and 6.8.2.

The directivity of each sound source is also an important consideration since most real sound sources do not emit sound in all directions equally. It's also the case that sound radiation patterns of acoustic sources are frequency-dependent. Low frequencies generally radiate in spherical patterns and high frequencies generally radiate more directionally. In musical instrument sound level measurements, Meyer (2009a)[1] found no spherical radiation patterns above 500 Hz. Musical instruments can have complex directional patterns that are sometimes dependent on the specific notes being played and vary from instrument to instrument. Some approximation of sound source directivity will help produce a more realistic auditory scene, especially for speech. For example, when a person turns away while they are talking, we still hear their voice but since high frequencies are radiated mostly straight out of their mouth rather than in all directions, we hear most of the high frequency energy only after it has reflected off nearby walls. Reflections in speech impact clarity and intelligibility depending on their timing and density. Since low frequencies are more omnidirectional in most sound sources, we hear those as both direct and reflected sound, but low frequencies do not help with speech clarity or intelligibility. We discuss source directivity further in Section 6.5.2.

3.5.2 Realism and Plausibility of Acoustic Simulation

Geometrical models of acoustic spaces can vary in their level of complexity and detail, which determine the computational resources required to render the virtual acoustics of a VR scene. Can we reduce the amount of detail in an acoustical model to save computational resources? Is it necessary that the smallest details of a space are included in a model to create a sense of immersion? How much detail is enough for plausibility? Research such as that by Vorländer (2020a) and Fichna, van de Par, and Ewert (2023) indicates that VR sound need not incorporate perfect geometrical models of a given space. It appears that the accuracy of early reflections is less important provided that the diffuse late reverberation is appropriate for the size of the virtual space. If the virtual acoustics are *perceptually* correct rather than physically correct, developers can save on processing power needed to render the sound, allowing content creators to approximate the acoustics of a scene using acoustical models with reduced computational complexity. If the sound design seems plausible, there is a good chance users will feel transported to and immersed in the virtual world.

While complex and precise acoustical models may not be necessary to create a sense of immersion and presence, that does not always mean that sound design can be simplified. The complexity of film sound design, for example, may not match the complexity of a scene, and it seems reasonable to assume that the same would hold true for VR sound. As the legendary film sound designer Walter Murch has written, seemingly simple film scenes can sometimes require many layers of sound to be effective, and conversely sometimes complex scenes can be accomplished with just a few key sounds. As he writes, "it can be simple to be complex, and complicated to be simple" (Murch 2005). His comment refers to a sound designer's creative process to produce plausible sonic environments that may employ layers of sounds, reverberation, and delays. Film sound designers may not rely on precise acoustical models to generate an appropriate sense of space for each scene, but they often work to achieve perceptually correct or hyper-realistic (exaggerated) scenes through sound layering, which we discuss further in Section 5.4.4. A simple scene, such as a door closing, for example, can require complex layering of sounds to provide the viewer with an appropriate impression of the event. In fact, realistically recorded sounds often may not sound realistic when paired with film (perhaps because the sound is not dramatic enough) and often need to be augmented and made hyper-realistic with additional layers of sound and signal processing just to make them sound realistic.

To produce effective and immersive audio for VR film or games that situate a user within a virtual space, early reflections and reverberation should match the visual dimensions of the virtual space and the user's location and orientation in that space. The geometry and surface materials within a room determine the early reflection patterns and reverberation. Therefore, a small, carpeted living room, for example, will have much different early reflection patterns and reverberation than those of a gymnasium or parking garage, spaces that have much larger volumes and more hard, reflective surfaces.

Standing waves or room modes, which we discussed in Section 3.3, are prominent in small rooms and can be approximated quite easily in the virtual domain. Artificially generated first-order room modes can help enable more accurate sound source localization for dynamically changing scenes in a five-channel loudspeaker system as Corey et al. (2001) reported, and it's not difficult to imagine that it would also be important for VR sound when the scene involves a small space.

3.5.3 Artificial Reverberation

The most widely used type of digital reverberation, sometimes referred to as algorithmic reverb, simulates sound decay in a space, also known as the reverberation tail, without using precise geometric models. While reverberation algorithms often include early reflection patterns based on models of real spaces, the reverberation tail itself is usually an approximation that, while realistic sounding, requires less processing power than precise geometric models computed in real time. Instead of tracing the paths of each reflection through a modeled space to generate the reverberation tail up to some very high order of reflections, conventional parametric digital reverberation algorithms use networks of delays with feedback connections to create very realistic sounding reverberation. Algorithmic reverberation approximates the way sound decays in a real space. In these recirculating delay networks, sound is attenuated slightly each time it goes through a feedback cycle. The result is a gradually decaying sound level that approximates reverberation decay in a real acoustic space. In sound design, artificial reverberation parameters such as reverberation time, predelay time, and early reflection patterns should be consistent with the size of the virtual space.

Another way to produce artificial reverberation is with an impulse response of an acoustic space. An impulse is a very short but very loud sound (such as a balloon popping), and an impulse response captures the sound of that impulse in a room. In other words, an impulse response is a snapshot of the way sound bounces around a room and gradually dies out. By merging an impulse response with a recorded track, the track can sound like it was recorded in the acoustic space from which the impulse response was taken. The mathematical process of combining an impulse response with an audio signal is known as convolution, so plugins that use impulse responses of rooms are often called IR reverb (for impulse response) or convolution reverb. The main advantage of convolution reverb plugins is that they can sound much more realistic than algorithmic reverb plugins because they use actual reflection patterns and reverberation decay curves of real acoustic spaces, whereas algorithmic reverbs are approximations of the behavior of sound in real spaces created with delays, filters, and feedback loops. The main problem with convolution reverb is that to change the size or character of a virtual acoustic room, a new impulse response must be loaded in. Most convolution reverb plugins allow some control over parameters such as decay time, but they are not as flexible as algorithmic reverbs.

3.5.4 Distance Perception

In natural listening, distance perception is determined by several factors, which can be incorporated into the sound design for virtual spaces:

- Loudness, at least for familiar sound sources. For unfamiliar sources, we don't know how loud they would normally be so we can't use loudness to judge how close they are. For moving sound sources, loudness changes as a source moves closer or further away.

- Initial time delay between the direct sound and first reflections. Sounds that are closer will likely have a greater time delay between the direct sound and first reflection.
- Ratio of direct and indirect sound. The higher the level of indirect sound relative to the direct sound, the further away the sound is.
- Motion parallax. Objects moving at a constant speed will appear to be moving a greater amount if they are closer to us than if they were at a greater distance. Doppler effect should be added for sound moving quickly, such as we hear with emergency vehicle sirens that rise in pitch if they are approaching us and drop in pitch if moving away from us.
- High frequency damping. High frequencies are absorbed by air before other frequencies, but this is temperature and humidity dependent. The further away a sound source is located, the more that high frequencies will be reduced.

While reflections and reverberation influence our perception of sound source distance, in a virtual acoustic environment distance perception is determined primarily by direct sound level. As Neidhardt (2023) reported, when a listener is within the critical distance of a sound source in a virtual room, a sound designer can create a plausible impression of changing sound source distance by changing only the direct sound level. Even if early reflection patterns and reverberation remain constant, and therefore independent of listener location, they have no influence on perceived sound source distance. For directional sound sources in which the direct sound is at a low level, early reflection patterns related to listener and source location are much more important.

Although reflections may not be necessary for sound source distance perception, they can aid in sound localization and reinforce a source's perceived location. Listening tests by Corey and Woszczyk (2002) indicate that fourth-order early reflection patterns generated using an image model of a rectangular room reinforce direct sound locations and are especially helpful for localization of lateral sources in a 5-channel loudspeaker system. Panning sounds to the side in a 5-channel playback system is inherently difficult because of the distance between front and surround loudspeakers and the fact that we have one ear on each side of the head, but added reflections appear to help stabilize laterally panned sources. The research also indicated that listeners were able to localize sounds accurately even when the direct sound was removed and only the fourth-order image model reflections remained.

3.6 Listening to an Acoustic Space

If we visit a large, unfamiliar space with hard, reflective surfaces such as a large lobby, a concert hall, or a large house of worship with high ceilings and large flat and hard surfaces, we may immediately notice acoustic qualities of the space, such as a long reverberation decay time and prominent reflections or echoes. The reverberation time and the arrival of reflections from all around us might help us feel immersed in sound. At the other end of the reverberation time scale, there are specially equipped laboratory rooms with no reflected sound and no reverberation, rooms known as anechoic chambers. An anechoic chamber is a room in which all the surfaces, walls, floor, and ceiling, are covered with deep, highly sound absorbent fiberglass wedges that eliminate or absorb all (or almost all) sound across a wide frequency range. Therefore, any echoes or reflected sound that are present in a normal room are completely absent in an anechoic chamber. Anechoic chambers usually have a hard wire mesh floor to walk on so that sound absorption wedges can be installed on the floor below the wire mesh. if you have ever had the chance to visit an anechoic chamber, you know that it can be a jarring experience. In fact, some people experience nausea when they enter an anechoic chamber. These kinds of experiences highlight the importance of our hearing for our sense of well-being and the fact that the auditory system is always monitoring cues from our environment, whether we are consciously paying attention to them or not.

We rely on the spatialization of direct and reflected sounds to determine our orientation and location in a space and the location of people and sound-producing objects around us. One of the big challenges with audio for VR is to generate an auditory illusion so that users perceive a virtual space as though it is real, that the auditory impression matches the visual impression, and that a VR user's attention is not drawn to the audio because of an improperly rendered auditory image. It is possible to fool the ears with sophisticated HRTF filtering, head tracking, and room acoustics simulation or reproduction, but it is also remarkably easy to fail at creating the illusion because of users' lived experiences in real acoustic spaces. If an auditory illusion fails in VR, a user's VR experience can deteriorate, and the sense of immersion can be lost.

3.7 Three Psychological Levels of Hearing

We might take for granted the natural spatialization of sound in real space because we often rely on auditory impressions of our surroundings automatically and without much conscious thought as we carry on conversations and interact with people and objects around us. According to D. A. Ramsdell (1970), normal hearing occurs at three psychological levels: symbolic, signal, and auditory background. It is valuable to consider these three levels that Ramsdell described as we design and mix sound for VR.

On the social level, we use sound to communicate with each other through language. Ramsdell refers to this first level as the *symbolic* level since words are symbols for objects and concepts, and language is symbolic in nature.

The second is the *signal*, or *warning*, level. Part of the role of the auditory system is to help keep us safe, alive, and out of harm's way. We respond involuntarily to unexpectedly and suddenly loud sounds, sometimes with a startle reflex in which we may move quickly and suddenly. Focused on survival, the second level allows us to adjust our everyday activities to navigate safely through the world. Because the visual sense is limited to our field of view, which is narrow compared to our aural sense, we can use our hearing to be alert of dangers that may originate outside our field of view. Our sense of hearing can also magnify the urgency of responding to danger that we may perceive visually, especially for particularly loud or timbrally unusual sounds. The *signal* component of the second level encompasses sounds which may elicit emotional responses to aesthetic experiences, such as music or nature sounds. It is possible that the calming feeling that may be evoked by nature sounds may, in part, be due to the spatial characteristics of the sounds. For example, waves crashing into a shoreline can create a wide spatial auditory extent, especially if we are positioned close to the shore and facing the water. Likewise, being in a forest with sounds such as birdsongs and leaves rustling all around us might create a sense of immersion and envelopment, due to the spatialization of reflected sound and reverberation. Perhaps feeling a sense of immersion within a sound field is one reason it can have a calming effect.

Ramsdell referred to the third level as the *auditory background* of all daily living and refers to it as the primitive level of hearing. Ramsdell (1970, pp. 437-438) writes: "At the primitive level of hearing, we react to the changing background sounds of the world around us *without being aware that we hear them.*" In the third level of hearing, background sounds in our environment, such as the noise of distant traffic, hum of air-handling fans, and rustling of leaves, contribute to our sense that we are part of the living world, even when we don't actively pay attention to them. It seems that reflected sound and reverberation are included in the third level of hearing, since we may react to them but not be consciously aware that we are hearing them. Reflections and reverberation are therefore crucial components of a soundscape, whether virtual or real. As sound designers and audio engineers we need to pay close attention to them to create convincing virtual auditory environments. Our auditory systems are also continuously monitoring incoming sound to judge the size of the space we are in and how we are situated in it through our perception of reflected sound. Early reflections and reverberation provide important cues to help us judge the size of a room and whether the "sound" of a room matches our visual impression. If we are outside, the distance to nearby houses or buildings determines the time delay of reflected sound. Reflected sound is an important indicator of our perception of a room's size if we are indoors. Reflected sound is also important in our perception of the distance to nearby buildings, trees, or other large objects if we are outside. While we may not pay much attention to our sense of a room's size in our everyday activities in familiar spaces, a familiar room can sound very different when furniture is removed or if carpet is added or removed. Our perceptual systems are constantly monitoring for changes to our environment. It is much more efficient use of our energy for our perceptual systems to focus mainly on changes as opposed to paying attention to every detail of a very familiar space. In unfamiliar spaces, our perceptual systems are operating with heightened attention.

Perception of three-dimensional spatial sound is a natural, everyday human experience, and, in contrast to vision, the human auditory system is omnidirectional. We can perceive sound arriving from any direction at any time. Spatial audio reproduced over headphones or loudspeakers seeks to recreate the acoustic experience of three-dimensional sound, whether through the recording and reproduction of sound in an acoustic space or through the synthesis and rendering of novel sound sources or spaces. The spatialization of sound, like we experience in natural, acoustic listening, is essential to an immersive VR experience. Spatial audio reproduction allows listeners to localize sound sources from any position around them and gain a sense of not only an environment's size and shape but likewise that of sound sources through the reproduction or synthesis of both direct and indirect sound.

Note

1 Jürgen Meyer's classic book *Acoustics and the Performance of Music* (2009) provides detailed information about sound directivity of musical instruments and concert hall acoustics.

Chapter 4

Spatial Audio

Computational Implementation of Immersive Sound

Spatial audio, which is an intrinsic component of VR experiences, can be created using a variety of recording, processing, and rendering techniques. Immersive sound scenes take as their building blocks a range of spatial audio elements, from monophonic signals to fully immersive multichannel audio captured with large arrays of microphones. In this chapter, we explore how immersive sound, which we naturally experience in our daily lives, can be recreated by computational means. We start out with a conceptual discussion on the impact of spatial audio on presence and immersion in VR. We then examine the sound reproduction systems that are most often used for spatial audio. These include binaural audio, which employs headphones, and multichannel audio, which relies on loudspeakers. We discuss the theory behind these two broad categories of systems as well as trade-offs inherent in each. While there are numerous techniques for spatial audio recording and reproduction, we discuss the ones that are most commonly used for VR production and broader immersive audio applications. Finally, we explore spatial audio mixing and rendering techniques, particularly focusing on Ambisonics and binaural audio.

4.1 Spatial Audio and Immersion

Spatial audio plays a crucial role in virtual reality because it can enhance the sense of immersion and presence that users experience in a virtual environment. In this chapter, we discuss sound reproduction techniques and paradigms, as well as microphone techniques that have been developed to capture spatial sound for later reproduction. Before getting into specific techniques and technologies for spatial audio, we will touch on some higher-level concepts related to overarching goals and challenges of VR, and to summarize some relevant research which can guide VR creators' practical design decisions.

In a quantitative survey of users consuming VR, Mütterlein (2018) investigated users' VR experience and found that the key features of VR, namely immersion, presence, and interactivity, are interrelated. Specifically, he found that interactivity contributes to users' sense of presence and immersion, that presence is crucial for immersion, and that immersion has a direct influence on satisfaction. From a practical point of view, we can extrapolate from the findings that to create a sense of immersion, it is important to include spatial audio features that support and enhance interactivity and that make users feel present in virtual environments.

The question may arise: how do we define immersion, presence, and interactivity in the context of VR? While some authors describe them in interrelated ways, Slater and Wilbur (1997) distinguish between immersion and presence. They write that immersion is

DOI: 10.4324/9781003048022-4

objective and quantifiable in contrast to presence, which is a state of consciousness. Immersion depends on a VR technology's capability to deliver an inclusive, extensive, surrounding, and vivid sense of reality to the user, described briefly as follows:

- Inclusive: How well is the user closed off or isolated from the real, physical world?
- Extensive: To what extent are all the senses engaged?
- Surrounding: How wide is the field of view within the virtual environment?
- Vivid: How accurately is the virtual scene represented? This aspect typically corresponds to video resolution.

Agrawal et al. (2020) write that

> immersion is a phenomenon experienced by an individual when they are in a state of deep mental involvement in which their cognitive processes (with or without sensory stimulation) cause a shift in their attentional state such that one may experience disassociation from the awareness of the physical world.

In terms of audio and acoustics, immersion usually refers to the "psychological sensation of being surrounded by specific sound sources as well as ambient sound" as Wenzel, Begault, and Godfroy-Cooper (2017) write. While early VR systems were developed for 3D vision without sound, spatial audio with real-time interaction and processing is essential for creating a sense of immersion and presence in VR. Expanding on Mütterlein's finding that immersion, presence, and interactivity are interrelated, Vorländer (2020b) writes that a VR environment requires the following to create a sense of immersion:

- interactivity, instantaneous responsiveness, and externalization of sound (in contrast to in-head localization of sound),
- smooth audio and video,
- synchronized audio and video,
- matching intensity of audio and video, and
- matching spatial sounds with visual spatial impression.

Work by Woszczyk, Bech, and Hansen (1995), which investigated interactions between seeing and hearing in the context of a home theater system, is consistent with Vorländer. They report that hearing and vision are interdependent, and that audio and video signals should be synchronized, have spatial coincidence (that is, they should be co-located), have congruent movement, have balance between picture size and loudness of sound, and match in quality.

Shivappa et al. (2016) similarly report that compelling immersive audio over headphones requires accurate localization of sounds in all directions and dynamic binaural spatialization or head tracking. It also requires that any latency between motion and sound should be undetectable by users. In listening tests, Potter, Cvetkovic, and De Sena (2022) found that adding reverberation has the same effect on immersion as increasing the video resolution five-fold from 0.5 megapixels to 2.5 megapixels per eye. Again, we find more evidence of the power of audio in a visually dominated medium.

While presence is sometimes described in relation to immersion, it is a distinct concept for many researchers. When a user interacts with a virtual environment, to what degree can they momentarily "forget" that the visuals are originating from a video display and the sounds are being reproduced over headphones? Michael Abrash (2014), chief scientist at Meta Reality Labs, argues that presence is the sense of being someplace other than one's physical location while in VR. Lombard and Ditton (1997) define presence as "the perceptual illusion of nonmediation" and an "illusion that a mediated experience is not mediated." IJsselsteijn et al. (2000) describe presence as "the sense of 'being there' in a mediated environment." They state that audio and video fidelity along with a tight connection between a user's motion and the corresponding updates in the visual and auditory displays, usually provided by head tracking and other motion sensors, are most important for creating a sense of presence. Presence is also affected by the relationship between direct sound, indirect sound (reflections and reverberation), and listeners' perceptions of the sound, as Kelly, Woszczyk, and King (2020) report in their literature review on the topic.

While there may be a tendency to either ignore background sounds or attempt to remove them, Gilkey and Weisenberger (1995) argue that background or incidental sound, referred to as "room tone" by audio engineers, is crucial for creating a sense of presence. Room tone is the sound of a room in silence, consisting of quiet background sounds such as air handling noise, the hum of lights and appliances, a clock ticking, or wind or traffic noise from outside. Room tone is present everywhere in our daily lives even though we ignore it most of the time. As quiet as a room may be, room tone helps us feel like we are situated within a space even if we are not consciously paying attention to it. Take a moment right now and listen to the ambient sound and room tone in your environment.

Film sound designers leverage room tone because scenes without ambient sound can feel unnatural and unsettling. Although we may not pay conscious attention to quiet ambient sounds, they can reinforce visual scenes and help create a sense of presence. In their research on the role of spatial audio in extended reality applications, Kailas and Tiwari (2021) found that listeners felt discomfort with total silence. In a study by Lindquist et al. (2020), participants were presented with digital animations of five different outdoor environments displayed through a conventional screen and a VR head-mounted display, each with three different sound conditions: no sound, ambient sound, and detailed sound. They found that the use of ambient and realistic sounds increased participants' perceived realism, with a more pronounced effect in VR.

Presence might be negatively affected by inconsistencies in the sound environment, for instance, if there are particularly jarring combinations of natural and synthetic sounds, or if indoor and outdoor sounds are intermixed (Larsson et al. 2010). While sound designers often combine natural and synthetic sounds with a high degree of success, inconsistencies in sound quality can distract users when timbres and amplitude envelopes are not carefully shaped. Sound designers should be aware of abnormalities that could arise when rendering motion trajectories. For example, if a VR scene uses the sound of a slow-moving car but displays a speeding car, the sound will not match the video. Sound design choices should be informed by their ecological validity, that is, the requirement to provide auditory cues that conform to our expectations of sound in real life. If we achieve all the conditions to create a perfect sense of presence in a virtual environment, users may be more likely to ignore the presence of the sound reproduction system and video display. Users may perceive a perceptually accurate VR environment to be as convincing as real life because the sense of presence this environment instills can make them feel as though they can see and hear as well as they can in real life.

Real-time interactivity is one of the features of the VR medium that distinguishes it from traditional, fixed media such as movies played back over conventional surround sound systems. Interactivity in a VR context can be as minimal as being able to turn one's head and look around. Even in linear formats such as audio books, listeners experience a greater level of immersion into the story when the storytelling involves real-time spatial audio interaction based on head-tracked binaural audio compared to monophonic and static binaural rendering (Geronazzo et al. 2019). In Chapter 6, we explore various concepts and techniques for implementing real-time interactivity in VR audio.

Brandenburg et al. (2023) list six main psychoacoustic cues that contribute to a plausibly realistic binaural rendering of an immersive virtual acoustic scene:

1. Interaural time and level differences (ITD and ILD): The human auditory system relies on time delays (mainly for low frequencies) and level differences (mainly for high frequencies) of sound arriving at the ears to localize sounds.
2. Visual cues: The visual location of a source can help solidify its perceived auditory location. We tend to localize a sound to its visual location even if the sound is not arriving from the same location. The phenomenon is called the ventriloquist effect because ventriloquists rely on this phenomenon to make it look like their voice is coming from a puppet. Visual cues also help with estimating a sound's distance.
3. Head-related transfer function (HRTF): The outer ears (pinnae), head, and shoulders filter incoming sound before it reaches the ear drum, and this filtering provides important cues for sound localization. Due to variations from person to person in head and ear shape and size, it was thought that individualized HRTFs were essential for the most accurate sound localization and externalization. Newer research indicates that generic HRTFs can be used effectively if other cues, such as dynamically updated HRTFs linked to head tracking, are included.
4. Familiarity with an acoustic environment: When we recognize an acoustic environment, it results in better plausibility and better externalization.
5. Room congruence: Although more applicable in an augmented reality (AR) context, simulated virtual acoustics should plausibly match the room in which a listener is located for the most effective externalization of sounds.
6. Motion: We have known for some time that when we update HRTFs based on head rotation, it leads to better sound externalization and localization. What has become clearer recently is that a user's movement through a space, also known as vection, is an essential cue for immersion.

As these researchers also point out, not only is it important for these psychoacoustic cues to be present but it's also important that unnatural cues not be present, since unnatural cues can destroy the plausibility of a binaurally rendered scene, negatively impacting immersion.

4.2 Spatial Audio Reproduction

Spatial audio, also called immersive audio or 3D audio, generally refers to sound reproduction using headphones or multichannel loudspeaker arrays such as stereo, quadrophony, surround sound (which includes formats such as 5.1-, 7.1-, 9.1-, 10.2-, and 22.2-channel audio systems), Ambisonics, Auro-3D, and wavefield synthesis (Nicol 2020).

Spatial audio reproduced binaurally over headphones allows VR designers to generate sound from any direction, with dynamic audio rendering and processing based on the location, orientation, and movement of users and sound sources in a VR environment. The combination of binaural audio and dynamic processing creates an immersive sonic experience akin to real-world listening. While most VR applications rely on headphones alone, subwoofers can be used to supplement the audio in the low frequency range, which can be important for whole-body vibration and the perception of bass level (Simon et al. 2009). Low frequencies from subwoofers can easily pass through closed headphones and can enhance the VR experience when those frequencies need to be emphasized. Despite the widespread use of headphones for VR, multichannel loudspeaker systems remain essential for immersive sound in movies, games, and music. However, while binaural audio can generate sound from any direction, most multichannel loudspeaker formats do not include loudspeakers below ear level and as such do not generate truly 3D audio. Nevertheless, this chapter will cover concepts and techniques that pertain to both headphone and loudspeaker delivery methods.

4.2.1 Headphone Sound Reproduction

While there may be no perfect sound reproduction technology, headphones have advantages over loudspeakers, especially in the context of VR. One of these advantages is their portability, which allows a listener to experience immersive sound wherever they happen to be located. In terms of interactive immersive sound and control of virtual acoustics, headphones allow the theoretical ability to recreate any soundfield at a listener's ears, making them ideal for VR applications. Furthermore, headphones with closed backs provide some acoustic isolation, which can help with presence.

As we discussed in Section 3.4, the term binaural audio usually refers to headphone reproduction of two audio signals (left and right channels) that incorporate interaural time differences (ITD), interaural level differences (ILD), and filtering by the outer ears, typically by processing the audio with the appropriate HRTFs for each virtual sound source location. The term binaural *audio* is distinct from the term binaural *hearing*, which is the natural listening experience we have in everyday life without headphones. One of the main features of binaural audio is that it can produce externalized and localized sound, approximating natural hearing using headphone reproduction. Conventional stereo recordings played back over headphones are not considered to be binaural because they do not integrate cues for sound localization and externalization, and they have not been processed with HRTFs. When we listen to traditional stereo audio over headphones, we tend to localize center-panned sounds in the middle of the head, a phenomenon known as intra-cranial (inside-the-head) localization or lateralization. Binaural audio, on the other hand, can mitigate this effect by externalizing sounds, mimicking the natural acoustic listening experience. Because binaural audio reproduction over headphones is unaffected by listening room acoustics, it allows full control over a VR environment's virtual acoustics, providing a consistent listening experience regardless of the acoustic conditions of the room in which the user is located.

If binaural audio reproduced over headphones can recreate sound waves exactly as they would have arrived at the ears in normal listening without headphones, then theoretically it is possible for binaural audio reproduced over headphones to be indistinguishable from natural hearing without headphones. Accurate binaural audio can create the impression that a sound source is originating from somewhere other than the headphones one is wearing, which means that sounds are all externalized. With immersive and interactive VR audio, the aim is for listeners to forget that they are wearing headphones.

We can create binaural audio signals using one of two main methods:

1 by recording an acoustic sound source with a binaural or dummy head, or
2 by processing mono signals—whether those mono signals are prerecorded or synthesized in real-time—through head-related transfer functions (HRTFs).

We can make binaural recordings with a device called a dummy head, also known as a binaural head or head and torso simulator (HATS), which is a life-size replica of a human head with life-like ear replicas that have omnidirectional microphones placed in the ear canals. Perhaps the most well-known commercially available dummy head is the Neumann KU 100, but there are other options such as the less expensive binaural microphone from 3Dio, which consists of only two spaced outer ears without the head replica. It is also possible to build a dummy head with a mannequin head and a pair of microphones, or to place miniature microphones at the entrances to one's own ear canals to make a binaural recording. Binaural recordings can be remarkably realistic, and when played back over headphones, listeners can believe they are physically situated in the space in which the recording was made.

As Algazi, Duda, and Thompson (2004) report, binaural recordings are not without their problems. The first issue is that the outer-ear shapes and sizes on a dummy head may be different than those of the listeners. These differences can cause localization errors, predominantly with sound source elevation but also angular location (azimuth).

The second issue is that the auditory field reproduced over headphones turns with a listener when they turn their head. In other words, the sounds are head-locked and thus move with head movements. Head-locked sound is, of course, common for headphone listening outside of VR experiences, but it is at odds with natural listening in which sound source locations are stable and independent of head movements.

The third issue is that we already experience front-back localization errors in natural listening due to the cone of confusion, which we discussed in Section 3.2.4. We described how we use small head rotations, often subconsciously, to resolve sound source location when we encounter localization errors or ambiguities in natural hearing. A slight turn of the head changes the interaural differences, which causes sound to arrive sooner and louder in one ear and later and quieter in the other ear. The resulting change in interaural differences helps resolve source location ambiguity. However, such head turns won't resolve ambiguities encountered in head-locked binaural audio since head movements are not tracked.

Because head movements are crucial for sound localization, head tracking in VR audio plays a critical role to help users resolve front-back and up-down confusions when listening binaurally over headphones. In a virtual environment, we want sound sources to be anchored in their respective locations, just like in natural listening, where sound sources do not move when we turn our heads, but stay fixed at their physical location. Virtual sound sources should not move when we turn our heads, unless this serves as a particular purpose, such as delivering non-diegetic dialogue or creating the feeling of headphone listening for selected sounds. If we want sounds to remain locked in place in VR, the system needs to have head tracking to control the headphone audio. A head-tracked binaural audio system monitors the position, orientation, and movement of a user's head, which allows the system to dynamically change the audio so that it mimics the way we perceive sound in the real world. For example, if a user hears a virtual sound to their right and turns their head away from the sound (to the left, in this case), the user will perceive the sound as originating from behind.

While modern VR systems already incorporate head tracking, this technology has been increasingly integrated into consumer headphones and ear buds, such as Apple AirPods Pro and AirPods Max, Beats Studio Pro, LG TONE Free T90, Samsung Galaxy Buds Pro, and Sony WH-1000XM5 and WH-1000XM6. Head tracking provides stable sound images for listening to spatial audio music or watching movies even when not specifically in a VR environment. Compatible headphones transmit a listener's real-time head acceleration and rotation rate data to a connected mobile device. As a result, listeners can experience head-tracked spatial audio when listening to music or watching video on a compatible device. Practically, this means that a spatial sound image stays relatively stable—ideally locked to the location of a screen—as a listener moves their head.

Binaural audio rendered using head-related transfer functions (HRTFs) offers the greatest potential for realistic control of sound source position and spatialization, but with some tradeoffs. As we discussed in Section 3.4, the main issue with HRTFs is that no one set works equally well for everyone. However, head tracking may be the most important technique to optimize the spatial audio experience, since it improves sound localization accuracy and provides externalization, even when using generic, non-individualized HRTFs (Sunder et al. 2023). Head tracking allows a virtualized soundfield to update dynamically and respond in real time to head movement, allowing the experience to approximate natural listening.

In one experiment, Rummukainen, Robotham, and Habets (2021) found that listeners favored individualized HRTFs for static sound localization, consistent with research by others. In a second experiment, they compared individual HRTFs with generic HRTFs in a dynamic VR environment (with both head tracking and overall location tracking). They found that the generic HRTFs performed just as well as individualized HRTFs for users' perception of sound localization, timbre, and overall quality in a dynamic, tracked VR environment. The research seems to indicate that generic HRTFs are acceptable and that individual HRTFs may not be necessary when coupled with a motion-tracked, multimodal, and dynamic environment in VR.

A consideration when choosing a pair of headphones is whether they are closed-back or open-back. Closed-back headphones provide better passive sound isolation than open-back headphones, although they are not as effective in that regard as headphones that incorporate active noise cancellation. Open-back headphones provide almost no sound isolation which means that users can hear external sounds, which can be a positive or negative feature depending on the application. The environmental awareness afforded by open headphones may be helpful for users to hear sounds around them, such as strategically placed loudspeakers for a VR experience. Some headphones and earbuds allow listeners to switch between active noise-canceling mode, which significantly attenuates surrounding acoustic sounds, and transparency mode, which allows users to hear these sounds while wearing the headphones.

4.2.2 Loudspeaker Sound Reproduction

Many people have experienced spatial audio in the form of surround sound reproduced over loudspeakers, primarily through movies in commercial cinemas and home theaters. Loudspeakers have an advantage over headphones in that hundreds of people can experience spatial audio together while watching a movie without the need to wear individual headphones with dedicated feeds. Movie goers get to share the experience and hear reactions from others in the theater while being immersed in sound. Loudspeakers are hard sound sources that remain in place unlike headphones which move with the listener. Loudspeakers around the listening area including height channels ensures stable sound source positioning without the need to use non-individualized HRTFs that may provide suboptimal localization for listeners. Likewise, neither HRTF processing nor head tracking is needed to experience stable immersive sound with loudspeaker reproduction.

Research indicates that sound reproduction over loudspeakers has perceptual advantages over heaphone sound reproduction. For example, Hong et al. (2019) found that the spatial reproduction quality of a 2D first-order Ambisonic octagonal loudspeaker array was higher than that of a first-order Ambisonic head-tracked binaural output reproduced over headphones. Their research indicates that perception of distance, direction, and elevation might not be as accurate in the first-order Ambisonic head-tracked headphone binaural system as it is with a multichannel loudspeaker array. They did find, however, that the first-order Ambisonic head-tracked binaural and 2D first-order Ambisonic multichannel loudspeaker system performed equally well in the context of soundscape assessment and design using VR head-mounted displays.

To create optimal sound localization and immersion with multichannel loudspeakers there are a few practical factors to consider. One of the most important is listener location. The ideal listening position in a multichannel audio system is in the center of the array, equidistant from each loudspeaker, a position commonly known as the "sweet spot." Listeners located outside of the sweet spot do not have an ideal listening experience because they are closer to some loudspeakers than to others. Listening off-center affects the relative level and timing of sound from the loudspeakers which impairs sound localization and immersion. When a listener is seated off-center, sound from the closest loudspeaker(s) arrives slightly louder and a little sooner than sound from more distant loudspeakers. This is true even or perhaps especially when all loudspeakers are producing equal sound levels and timing. Since we rely on interaural level differences of a few decibels and interaural time differences less than 1 millisecond to localize sound, even a relatively small change in listening position affects the way we localize sound and experience immersion within a loudspeaker array. Unless listeners are in the sweet spot, listeners localize amplitude panned sound source towards the closest loudspeakers. The concept of a sweet spot is irrelevant to headphone listening since headphone locations are fixed at the ears and we do not need to be concerned with loudspeaker positioning.

Listener movement is another factor in loudspeaker reproduction. When listeners move around or turn their heads while listening to a loudspeaker array, it affects the way they localize sound from the loudspeakers and impacts their sense of immersion. This has to do with the way stereophony works. Listeners perceive a phantom image between front left and right loudspeakers when the loudspeakers emit the same signal without any time or level difference. The stereophony illusion is based on the "summing localization" effect, but it works only if the listener is facing the two loudspeakers and is equal distance from them. Turning to the side, listeners perceive phantom image placement to be shifted since the loudspeaker signals no longer sum at the two ears in the same way as when facing forward.

With loudspeaker arrays in large rooms, we can expect there to be multiple listeners in the same listening area, and sound from some loudspeakers may be blocked for some listeners if loudspeakers are at ear level. This issue can be mitigated if loudspeakers are located slightly above ear level and angled down, which creates a better chance of being heard equally by all listeners. Headphone listening on the other hand, ensures there is no chance that another listener or object will block sound from a loudspeaker.

Finally, another factor with loudspeakers is the cost. We often need four or more loudspeakers to generate immersive, multichannel audio. High-quality loudspeaker arrays cost significantly more than a pair of even the highest-quality headphones.

4.2.3 Setting the Scene: Controlling Sound Source Location, Distance, and Ambience

When reproducing audio, the most common way to control a sound's angular location (or azimuth) is with amplitude panning, also known as intensity panning. A panner changes a source's perceived phantom image location by controlling the interchannel amplitude difference signals sent to a pair of loudspeakers or headphone channels. However, amplitude panning does not work equally well for all locations within a multichannel surround system. Side and rear images can be particularly problematic mainly due to the nature of human hearing. In addition to challenges with phantom image control to the sides and rear, loudspeaker spacing can also cause issues with amplitude panning. Michael Gerzon (1985) wrote that loudspeakers spaced too far apart, that is, those that subtend an angle greater than 60°, reduce the quality of phantom images. Gerzon further wrote that loudspeaker pairs to the side of a listener create unstable phantom images, since the phantom image illusion breaks down, which has been confirmed by other researchers as well. For example, Martin et al. (1999a) conducted a series of listening tests to better understand the perception of phantom images created using interchannel amplitude and time differences between loudspeaker pairs within a five-channel surround sound configuration. After listening to amplitude-panned lateral phantom images from side loudspeaker pairs, participants reported that small front-back changes in listening position resulted in highly variable phantom image locations, consistent with Gerzon's report of unstable side phantom images.

In contrast to amplitude panning, time-delay panning, which varies an audio signal's relative timing to a pair of channels, is not nearly as effective as amplitude panning and therefore finds rare use, at least for loudspeaker reproduction. In an experiment that compared amplitude panning to time-delay panning over loudspeakers, Martin et al. (1999b) found that listeners were more uncertain about time-delay-panned phantom image locations compared to amplitude-panned phantom images. Listeners also reported that time-delay panning impacted phantom image "focus," meaning that they perceived static sound sources to originate from more than one direction when time-delay panning was used. Time-delay panning can also introduce unwanted timbral changes due to comb filtering. Crosstalk, a characteristic of loudspeaker listening in which sound from each loudspeaker reaches both ears, appears to be more of an issue for time-delay panning than for amplitude-based panning. Headphones are generally more suitable for time-delay panning since there is no crosstalk between the left and right channels. There is, however, an effective upper limit for interaural time difference panning, since the greatest interaural time difference in natural hearing is less than 800 microseconds or 0.8 milliseconds, which is about the time it takes for sound to travel from one ear to the other based on the average human ear spacing. With interaural time delays greater than that, listeners perceive that the two loudspeaker signals separate or widen. With time delays greater than about 30 milliseconds, listeners may hear an echo, depending on the sound source.

In addition to a sound source's angular location, we can use artificial or recorded reflections and reverberation to simulate an acoustic space. In a VR environment, it is especially useful to control perceived distance using audio (rather than with visuals alone), since it turns out that VR users tend to perceive objects as being closer than intended. The phenomenon to underestimate distance in a virtual world is known as *distance compression*. In their research on the effect of reverberation on VR distance compression, Huang et al. (2021) found that users associate shorter reverberation times with closer distance, and longer reverberation times with increased distance. Other researchers such as Bailey and Fazenda (2017) have also found that reverberation can increase a visual object's perceived distance and compensate for distance compression in VR. Thus, for more distant sources, increasing the reverberation time can help reduce this compression effect.

Research on distance perception in VR suggests that auditory cues for distance may override visual cues in some situations. In other words, users may judge distance based on auditory cues instead of visual cues. For a sound source at a given distance, and particularly for those in the near field, a perceptually accurate reverberation time is critical when compensating for distance compression. If reverberation time is too long, audio-visual segregation may increase. Segregation in this context means users have trouble unifying the auditory and visual sensory information, which is something we want to avoid.

Degrees of Freedom (DoF)

When a VR user changes the orientation of their head or moves to a different location, the audio and video need to update accordingly to change their virtual perspective. This requires the VR system to track the user's movements and use this data to control the audio and video. We describe the possible range of movement that a VR system can track by its "degrees of freedom." For example, in real life, buildings are rigid and fixed in space, so we say they have no degrees of freedom. A train on a track has one degree of freedom which allows it to move back and forth only.

A basic level of VR interactivity should allow users to orient their heads in any direction within a virtual space to see and hear objects or other users. This level of interactivity requires the system to have three degrees of freedom (3DoF). A 3DoF system can track a user's head rotation around three axes known as pitch, yaw, and roll. Those familiar with aviation or spaceflight may recognize these terms since they are commonly used to describe an aircraft's movement through space. Pitch rotation is nodding the head up or down to indicate "yes." The head rotates around the horizontal axis that connects the ears. Yaw rotation refers to turning the head from side to side to look left or right or to indicate "no," and it rotates around the axis that goes vertically through the neck. Roll refers to tilting the head from side to side, with rotation around the axis roughly centered at the nose. To optimize sound localization in real life or to resolve front-back localization confusions, we usually turn our heads (yaw), which changes interaural time and intensity differences dynamically in natural hearing. In VR, 3DoF head tracking allows audio signals to change dynamically in response to head orientation, which helps resolve localization ambiguities.

While head tracking allows a 3DoF VR system to update the audio and video dynamically based on head orientation, it does not track a user's location or movement through a space. A VR system needs six degrees of freedom (6DoF) for full movement tracking. A 6DoF system tracks not only head rotation (pitch, yaw, and roll), but also translational movement, that is, movement through a space in the front-back, left-right, and up-down dimensions. In other words, a 6DoF system allows users to walk around and move freely within a virtual space and this freedom of movement enhances the immersive experience.

4.3 Capturing Immersive Sound

In natural hearing without any loudspeakers or headphones, we localize sound sources and feel immersed in the direct and reflected sound that arrives from all directions with only the two acoustic signals from our left and right ears. How do we recreate a sense of localization and immersion with sound reproduced over headphones or loudspeakers? Microphones, whether we have only one or several dozen, are our main tools for capturing acoustic sounds. The primary challenge is to mix or transform those microphone signals in a way that will recreate an immersive experience for listeners and allow them to localize sounds, whether they are listening over headphones or loudspeakers. The situation is further complicated with VR audio because both the sound sources and the listener can move.

When we plan to record immersive sound for a VR experience, there are several points to consider, including:

- What auditory perspective do we want to create?
- Which microphones and techniques do we use to capture the chosen auditory perspective?
- What will be the extent of interactivity for users?
- Will any of the virtual elements (e.g., actors, objects) be moving?
- If the recording is part of a VR film production, how can we conceal microphones from the video recording?

Many immersive microphone techniques have been developed from or share underlying principles with conventional mono and stereo techniques. Furthermore, multiple mono and stereo microphone techniques can be combined in sophisticated ways to create immersive audio mixes. In this section, we will begin by reviewing some basic techniques used in mono and stereo recording and then move into immersive microphone techniques.

4.3.1 Spot and Stereo Microphone Techniques

A microphone's placement, proximity to a source, and orientation determines the quality of the captured sound. Where do we place microphones? How far away do we place microphones from sound sources? What are the tradeoffs between close and distant placement? How does a microphone's angle of orientation affect the sound? There are no hard and fast rules for microphone placement, so experimentation with microphone placement and orientation combined with careful listening to the resultant sound is often a good way to start.

In recording studio setups, engineers usually place at least one microphone close to each sound source, using a technique called spot-miking or close-miking. For VR applications, spot microphone signals are in many ways ideal because they can be rendered along with panning, equalization, mix levels, and artificial reverberation, all of which can be adjusted in real time as sources and listeners move within a space so that the most appropriate auditory perspective is created for the user. Spot microphones are usually placed close to a source and as such they provide more acoustic isolation from other sound sources, reverberation, and reflections than a more distant microphone. The isolation we get from spot microphones is highly useful in many recording situations since it is analogous to a spotlight in the visual domain which allows us to create focal points in the sound image. When recording many sound sources, spot miking allows us to adjust the relative levels of sound sources, whether these are in a traditional mix or in the context of a VR scene. Mixing and balancing microphone signals helps guide listeners and so they hear what is most appropriate at any given time. Mono microphone signals can be panned around a listener dynamically as the listener and other users and objects move within a scene.

While a recording made with a single, spot microphone offers flexibility for source placement within a 3D audio image, there are limitations inherent in a monophonic signal. We know from research on acoustic sound source radiation patterns that a single, closely positioned microphone is not likely to capture the entirety of an acoustic sound source's output unless the source is quite small. Most acoustic sound sources, including musical instruments, radiate sound in many directions, and sound sources may not always radiate sound equally in all directions for all frequencies. Although acoustic sound sources are usually not omnidirectional for all pitches, they do not typically have narrowly focused sound radiation patterns. Because of the potentially wide sound radiation patterns, some harmonics or frequency ranges of the sound may be radiated away from a spot microphone. In addition, the radiation patterns typically vary based on the note being played. For example, a woodwind instrument such as a clarinet radiates different pitches in different directions. Sound may radiate straight out of a clarinet's bell (its flared, open end) for a low-pitched note that requires all keys to be pressed, or it may radiate through the key holes for higher pitched notes in which most of the keys are open.

Sound source size also becomes an issue when recording with a single microphone. Larger sources such as waves on a beach, an orchestra, or a thunderstorm radiate sound from an expansive area and a single microphone cannot capture the full spatial extent of such a source. In other words, while mono sound recordings may be ideal for relatively small sound sources like the human voice, they may not convey an appropriate sense of width for large sound sources or large spaces.

Along these lines, no matter the size of the source, spot microphone signals on their own produce an artificially narrow sound that might distract a user and take them out of the experience. Reverberation captured by a spot microphone has a tunnel-like sound because direct sounds and reverberation originate from the same, narrow location in the mix. Due to the focused sound from a single microphone, it helps to add artificial reverberation or delays to widen a spot microphone's sound image. Artificial reverberation and early reflections also help integrate that signal into a mix and potentially give it a more natural sense of space.

We may choose close microphone placement to pick up less reverberation, since microphones placed very close to a source capture less indirect or reflected sound from the surrounding space. An acoustically dryer sound (less reverberation) may be a benefit or a drawback depending on the room acoustics and the need for reverberation in the final mix. If we record sound sources with close spot mics that attenuate early reflections and reverberation, we may need to use more artificial reverberation during post-production so that the sources do not sound artificially dry. When sources lack reverberation or a sense of space, users' attentions may be pulled out of the VR environment because they notice something unnatural about the sound.

Stereo microphone techniques, that is, those that use a pair of microphones, offer some benefits over mono, spot microphones in that regard. Stereo and multi-microphone arrays can produce a coherent sense of width and spaciousness for large sources and acoustic spaces, since they can capture not only a greater width but also the natural reverberation and early reflections of the acoustic space in which the sounds are located. Generally spot microphones are placed closer to sound sources while stereo techniques are placed further away to achieve more sound source blending and to capture some room reverberation.

Stereo microphone techniques can be divided into three main categories based on microphone capsule spacing: coincident, near-coincident, and spaced techniques. The most common coincident microphone techniques are Blumlein, mid/side (M/S), and XY. Each of these techniques uses a pair of directional microphones whose capsules are placed as close together as possible (coincident) and angled 90° apart, but the polar patterns of the microphones vary. XY uses two cardioid microphones, Blumlein uses two figure-8 microphones, and M/S uses a cardioid and a figure-8 microphone. XY and Blumlein microphone capsules are angled ±45° relative to the center of the sound source and their output signals are panned hard left and hard right. The M/S technique has a forward-facing cardioid or omnidirectional microphone (M) and a side-facing figure-8 microphone (S), but the signal panning is different than that of XY and Blumlein. The forward-facing M microphone is panned center, and the left-side-facing S microphone is split into two channels which are panned hard left and hard right. Importantly, the right-panned signal's polarity (or phase, often denoted on a mixer with the "Ø" symbol) is reversed. Without polarity reversal, the image is mono since the split S signals are otherwise identical.

Near-coincident stereo microphone techniques such as ORTF and NOS specify the use of cardioid microphones spaced apart slightly and angled outwards. ORTF microphone capsules should be spaced 17 cm (7 in.) and angled 110° apart. NOS specifies a slightly wider microphone capsule spacing of 30 cm (12 in.) but a slightly narrower angle of 90°.

One spaced microphone technique, known as an A-B pair, typically uses a pair of omnidirectional microphones spaced 40–60 cm apart (16–24 in.), although some engineers space them wider (e.g., 1.0–2.5 m; 40–98 in.) depending on the size of the ensemble or spacing of the sound sources. A Decca tree, used mainly for orchestra recording, includes a third omnidirectional microphone placed in the center of an A-B pair and about a 1 m (3 ft.) closer to the ensemble so that the three microphones form a triangle with a width of about 2 m (6 ft.) between the left and right microphones.

In terms of stereo imaging, the techniques differ notably due to differences in their spacing, orientation, and polar patterns. For example, the XY coincident technique produces a narrow stereo image that clumps the sound in the center of the phantom image, while the A-B spaced technique produces a wide stereo image with a great sense of space. The trade-off with an A-B pair is that while the sense of space and width may be strong, the locations of sources in the stereo image may be a little fuzzy. Near-coincident techniques such as ORTF and NOS generally have an even balance of energy across the stereo image along with clearer sound source locations.

While stereo microphone techniques have several benefits, they also have drawbacks. The main one is that they may not produce the ideal balance or mix of all sources. For large ensembles or sound sources, a stereo microphone pair may not pick up all sources equally well, in which case spot microphones can be used to augment sound sources that may be deficient in the stereo pair. Another issue is that stereo techniques, because of their more distant placement, often capture more room ambience than spot microphones, which can be undesirable in VR applications.

4.3.2 Correlation of Microphone Signals

The level of correlation between audio signals is relevant to spatial audio because it can affect the perceived width and spaciousness of a stereo or VR image. Correlation, simply put, is a measure of the similarity between two audio signals. At one end of the scale, two fully correlated signals are identical, therefore their correlation equals 1.0. When played back over stereo loudspeakers or headphones, two fully correlated signals produce the narrowest sound image, also known as monophonic sound.

When recording, it is helpful to keep in mind that microphone spacing affects correlation. Microphone capsules placed as close together as possible and pointed in the same direction will have a much higher level of correlation than those spaced further apart. Even if we pan these two microphone signals hard left and hard right, the resulting phantom image is going to be strongly monophonic and narrow because the microphone signals are so highly correlated.

At the other end of the correlation scale are negatively correlated signals where correlation equals -1.0. In this case the two signals are identical in every respect except one: polarity. Another way to say this is that one signal is inverted relative to the other. From a practical viewpoint, negatively correlated stereo signals, sometimes referred to as "out of phase," result in sound images that, when panned hard left and hard right, seem to originate from locations from outside of the stereo loudspeakers in an unnaturally wide sound image. The perceived positions of these wide sound images are highly sensitive to head movements, such that even small movements cause drastic shifts in the perceived locations of the sounds for most listeners. Because negatively correlated signals are of opposite polarity, panning them to the same location will cause them to cancel each other out due to destructive interference.

In the middle of the scale, decorrelated signals, where correlation equals 0.0, produce wider stereo images than correlated signals. One way to decrease the correlation of two microphone signals is to increase the capsule spacing since widely spaced microphones will be more decorrelated than coincident microphones pointing in the same direction. Two widely-spaced microphone signals panned hard left and hard right can result in two separate, distinct sound images, one at each loudspeaker with no center information, primarily because the two signals are decorrelated. An engineer may prefer a wide stereo image over a mono image in some situations because it can allow space for center-panned sounds. Decorrelated signals produce a stereo image with a "hole in the middle" of the phantom image. Another way to decorrelate directional microphone signals is to orient them in different directions. Two coincident cardioids facing opposite directions (angled 180° apart), for example, could potentially pick up substantially different acoustic signals resulting in low correlation.

Correlation usually also varies with frequency. For example, signals from a spaced pair of omnidirectional microphones may be decorrelated at high frequencies but correlated at low frequencies. The increased wavelength of low frequency sounds mitigates the decorrelation caused by the distance between microphone pairs. As a result, correlated low frequencies will appear in the center of a stereo image like a mono image, but decorrelated high frequencies will sound wider and more spacious.

4.3.3 Ambisonic Microphones and the Ambisonics Format

Ambisonics as a format provides the possibility to record, mix, and playback the full three dimensions of spatial audio. It includes microphone configurations for sound pickup as well as channel-based processing for post-production (mixing) and reproduction. In this section, we discuss Ambisonics first as a microphone technique. Later in this chapter, we will discuss Ambisonics from the mixing and sound reproduction point of view.

Traditional stereo microphone techniques, such as ORTF, X-Y, and Blumlein, capture sound in two dimensions, which is appropriate for two-channel audio reproduction. Drawing from principles in the M/S stereo technique, Ambisonic microphones capture not only distinct left/right spatial information but also front/back and height information, that is, the full sphere of incoming sound from all directions. While stereo microphone techniques also capture sound arriving from above and below, but they do not retain the spatial independence of the original soundfield in the way an Ambisonic microphone can. In a stereo microphone technique, sound arriving from above or below is collapsed into the horizontal plane. The result is that listeners are unable to hear the original elevation of sound sources. Likewise, sound arriving from the back is collapsed into sound arriving from the front with any stereo microphone array. An Ambisonic microphone, on the other hand, allows us to spatialize and separate sound from all directions, not only in the horizontal plane but also in the vertical plane. The other big advantage of Ambisonics as a recording technique or playback format is that it can be decoded to any number of loudspeakers or to binaural audio over headphones, so we do not need to make a new mix for each surround format.

Ambisonics is usually described in terms of its spatial resolution as denoted by its *order*. The higher the order, the more spatial resolution it provides, that is, the more precisely we can control sound source placement. The most common Ambisonic microphone is first order, which consists of four coincident subcardioid or cardioid microphone capsules arranged in a tetrahedron and pointing in these directions:

- Front, Left, and Up: FLU
- Front, Right, and Down: FRD
- Back, Left, and Down: BLD
- Back, Right, and Up: BRU

From these four signals, and the appropriate signal processing provided by Ambisonic plugins as we discuss below, we can reproduce a 360° soundfield. Tetrahedral, first-order Ambisonic microphones are manufactured by several different companies and include models such as Brahma Studio 4, Core Sound TetraMic (shown in Figure 4.1), Nevaton VR, Reynolds A-Type 4, RØDE NT-SF1 (aka "SoundField by RØDE"), Saramonic SR-VRMIC, and Sennheiser AMBEO VR 3D.

The four outputs taken straight out of a tetrahedral Ambisonic microphone as four discrete signals with no mixing or matrixing are known collectively as "A-Format." These outputs are indistinguishable from recordings made with four mono microphones. In other words, the signals do not amount to the intended spatial image by default. To achieve this, we must convert it to another format known as "B-Format," which provides a soundfield representation of the recording with the individual signals now correlated spatially. Once we have this B-format signal, we can decode it to any number of loudspeaker channels. There are a few plugins available that offer A-Format to B-Format conversion including dearVR AMBI MICRO, Audio Brewers ab Transcoder, and SoundField by RØDE.

Figure 4.1 A picture of the Core Sound TetraMic, which is a first-order, tetrahedral Ambisonic microphone. The image shows the four microphone capsules in black pointed in difference directions. (Image courtesy of Core Sound LLC.)

To retrieve the Ambisonic B-Format signal, the A-format outputs of the four capsules are combined into channels designated as W, X, Y, Z as follows:

- W = FLU + FRD + BLD + BRU
- X = FLU + FRD – BLD – BRU
- Y = FLU – FRD + BLD – BRU
- Z = FLU – FRD – BLD + BRU

What is interesting about these B-Format channels (W, X, Y, Z) is that they are equivalent to four virtual microphones with different polar patterns and orientations than a tetrahedral Ambisonic microphone. B-Format channels are equivalent to the following polar patterns and orientations:

- W: equivalent to an omnidirectional microphone that receives sound from all directions at equal intensity
- X: equivalent to a figure-8 microphone aimed forward
- Y: equivalent to a figure-8 microphone aimed to the left side
- Z: equivalent to a figure-8 microphone aimed up

In other words, B-Format is equivalent to three coincident M/S microphone pairs oriented to cover all three dimensions (front/back, left/right, up/down). Each of the B-Format channels (W, X, Y, Z) represents a different direction or dimension. We refer to these equivalents as *virtual* microphones since there are no actual omnidirectional or figure-8 microphone capsules present physically in an Ambisonic tetrahedral microphone. B-Format provides us with signals as though they originated from three figure-8 microphones (front/back, left/right, up/down) and an omnidirectional microphone. What this also means is that we can create first-order Ambisonic B-Format signals natively with an omnidirectional microphone and three figure-8 microphones (Lee 2021). In mathematical language, these four different microphone polar patterns, each pointing in a different direction, are called spherical harmonics. Figure 4.2 shows the first three orders of spherical harmonics. A tetrahedral mic, offering first-order Ambisonics, incorporates the first three harmonics, which coincide with the W, X, Y, and Z polar patterns that we described above. The higher the order, the more spherical harmonics, and the more precise panned sound source locations can be.

One important practical point to keep in mind is that microphone preamplifier gain settings for all outputs from an Ambisonic mic must be set to identical levels, otherwise the resulting spatial audio image will not accurately reflect the soundfield that was recorded. If the microphone gains are not matched and the gain differentials are not known, it will be impossible to calibrate the microphone signal levels to each other, and therefore the spatial imaging will be incorrect. Therefore, it is crucial to use identical and digitally-controllable microphone preamplifiers (rather than analog-controlled ones, which can be less precise) for each channel of an Ambisonic mic and ensure that the preamp gains are matched.

Furthermore, there cannot be any differences in timing or phase between the channels of an Ambisonic recording. Even very small time-delay differences between channels can destroy a spatial image, so we must take care to ensure that all recorded tracks from a

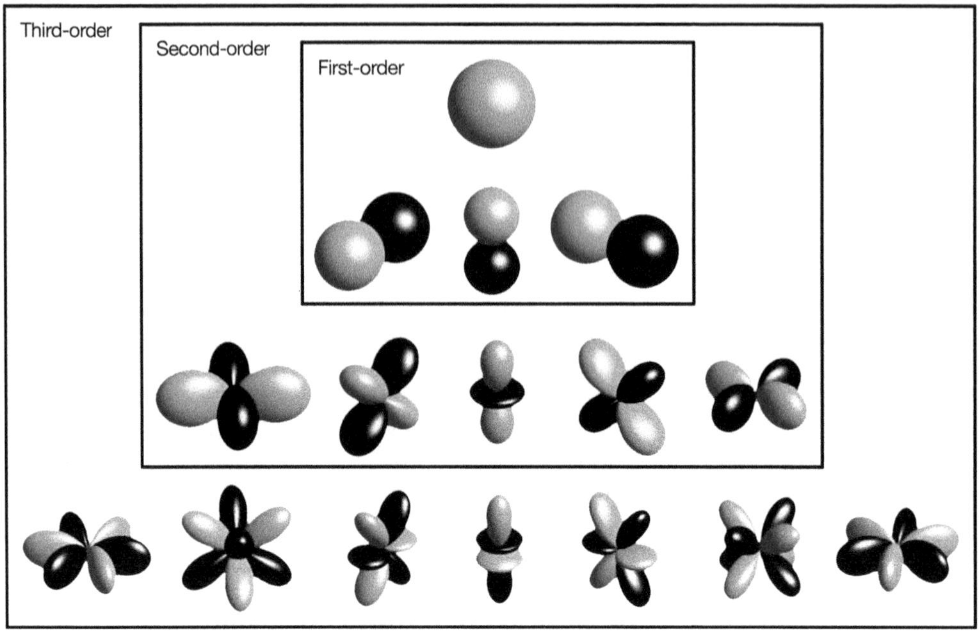

Figure 4.2 First-order Ambisonic spherical harmonics, which are shown in the smallest rectangle at the top, can be understood as polar patterns of four standard microphones: one omnidirectional microphone (top center), and three figure-8 microphones facing left/right, up/down, and front/back. Second-order Ambisonics, shown in the middle rectangle, includes the four first-order Ambisonic spherical harmonics and adds five more spherical harmonics. Third-order Ambisonic spherical harmonics, shown in the largest rectangle at the bottom, includes the nine spherical harmonics of the first two orders and adds seven more spherical harmonics. With more spherical harmonics, we have the possibility for more fine-grained control over sound source location.

tetrahedral microphone are processed as a group and that no individual microphone channel is shifted in time separate from the others.

For VR projects that involve real-world simulation, the engineer should note the microphone orientation and the sound pressure level (SPL) of the auditory environment being recorded. SPL authenticity can be essential to making a VR environment more realistic.

Equally important is the Ambisonic channel order to maintain the correct orientation of the recorded sound scene through any processing or audio file interchange that may occur. We need to maintain the Ambisonic B-Format channel order so that we can maintain correct sound imagining through all stages of production, post-production, and distribution. The two most common B-Format data exchange formats are known as FuMa and AmbiX. While the main differences between them are channel order and signal level, it is crucial to ensure that the encoder and decoder software format matches the format we have chosen to use, because FuMa and AmbiX are not interchangeable. Using an audio file in AmbiX format in a plugin that expects the FuMa format or vice versa can produce incorrect spatial rendering and localization errors.

FuMa, short for Furse-Malham, has a channel order of W, X, Y, Z, where the W channel is –3 dB relative to the other channels, but FuMa is impractical for higher order Ambisonics for a couple of reasons. First, it uses the WAV audio file format, which has a file size limit of 4 GB, and this is easily exceeded with higher order Ambisonic audio files. Second, the FuMa channel naming convention is letter-based and therefore impractical for orders higher than third or fourth order.

The AmbiX convention, developed at the University of Music and Performing Arts in Graz, Austria, is short for "Ambisonics Exchangeable" and is more widely used than FuMa. AmbiX specifies a different channel order and normalization scheme. The channel order of AmbiX is W, Y, Z, X, also known as Ambisonic Channel Number or ACN. The W channel in AmbiX is not attenuated like it is in FuMa. The normalization scheme used with AmbiX is known as SN3D, and it ensures that levels of all channels will not exceed the peak level of the W or omni component. AmbiX uses Apple's Core Audio Format (CAF) file format as a data container, which is unrestricted in terms of file size and number of channels. YouTube spatial audio and Waves Ambisonics plugins, for example, use the AmbiX format with ACN channel ordering and SN3D normalization. Other software such as Audiokinetic Wwise can import and play back B-Format audio in either FuMa (up to third order) or AmbiX (up to fifth order).

As we can see from the AmbiX channel order, the first two channels in this format (W and Y) are mid/side signals which can be decoded into two-channel stereo if needed. The W channel in B-Format corresponds to the mid channel, or omnidirectional microphone, and the Y channel in B-Format corresponds to the side channel, or side-facing figure-8 microphone.

While AmbiX is more commonly adopted, you may be required to convert your audio between FuMa and AmbiX, depending on audio software or delivery platform requirements. There are several plugins available that convert between these formats, including:

- Blue Ripple Sound O3A Core (free), Blue Ripple Sound O7A Core (paid)
- Penteo 360 (paid)
- SoundField by RØDE (free)
- Steinberg Cubase VST AmbiConverter Panel (paid; included with Cubase Pro)
- Waves B360 Ambisonics Encoder (paid)

The main advantage of Ambisonics is that when working with B-Format signals, we can play back the four channels over any loudspeaker layout or number of channels and even render the multichannel outputs to binaural audio for headphone listening. Contrast this with a non-Ambisonic multichannel mix made specifically for playback over a 5.1-channel audio reproduction system. A 5.1-channel mix does not translate easily to other loudspeaker layouts such as 7.1-, 9.1.4-, or 22.2-channel. With no direct channel-to-loudspeaker correspondence, it is not clear how to route audio signals in a 5.1-channel mix for playback over a system with more or fewer playback channels, requiring a separate mix for each loudspeaker system.

As we discuss further in Section 4.4.3, Ambisonic B-Format signals can be played back over almost any loudspeaker layout without requiring a different mix by providing the target loudspeaker layout to the Ambisonic decoder software. In other words, with a single Ambisonic microphone to capture sound in a space and the appropriate Ambisonic decoder, we can play a B-Format mix through almost any multichannel playback system configuration quickly and easily.

Although tetrahedral Ambisonic microphones can capture three-dimensional sound, they have a limited spatial resolution due to their limited number of channels. Lower spatial resolution means that sound locations are blurred, have a wider spread, and are less distinct. We can overcome the spatial resolution limitation by using a higher-order Ambisonic microphone (i.e., one with a greater number of physical microphone capsules), which provides more spherical harmonics and, therefore, increased spatial resolution. Second-order Ambisonic microphones, such as Brahma-8, Core Sound OctoMic, and Voyage Audio Spatial Mic, typically require eight capsules. When converted to B-Format, second-order Ambisonics has nine channels of audio that can be mixed to any loudspeaker layout, just like with first-order Ambisonics.

Higher orders of Ambisonics require more audio channels to encode the spatial audio information, but they provide increasingly higher spatial resolution. Again, the number of audio channels for each order of Ambisonics is independent of the playback system channel count. A first-order Ambisonic recording with 4 channels can be played back over a 5.1-channel system just as easily as a seventh-order Ambisonic recording with 64 channels can. The number of channels required to encode various orders of Ambisonics for a full 360° spherical soundfield is based on the formula $(N+1)^2$, where N represents the Ambisonic order. Therefore, the number of channels, for instance, required for Ambisonic orders one through ten would be as follows:

- First-order Ambisonics: 4 channels
- Second-order Ambisonics: 9 channels
- Third-order Ambisonics: 16 channels
- Fourth-order Ambisonics: 25 channels
- Fifth-order Ambisonics: 36 channels
- Sixth-order Ambisonics: 49 channels
- Seventh-order Ambisonics: 64 channels
- Tenth-order Ambisonics: 121 channels
- N^{th} order Ambisonics: $(N + 1)^2$ channels

Besides, the second-order Ambisonic microphones, there are also commercially available higher-order Ambisonic microphones, including the Zylia ZM-1 with 19 microphone capsules for third-order Ambisonics, and MH Acoustics em64 Eigenmike with 64 microphone capsule for sixth-order Ambisonics (shown in Figure 4.3).

While Ambisonics existed since long before the most recent wave of VR systems, it functions as an ideal spatial audio format for this medium. An Ambisonic microphone is a great match for 360° video, due in part to its practicality. Since an Ambisonic microphones does not favor a spatial direction, it matches well with the fully immersive nature of VR. Most traditional surround formats are biased towards the front, because in most cases they were developed for cinemas, in which the action is happening primarily on screen in front of the audience. Ambisonics, in contrast, captures and reproduces sound equally from all directions including above and below the listener. A tetrahedral Ambisonic microphone placed above or below a 360° camera remains out of the shot while capturing immersive audio at the location of the camera. The main shortcoming with a single

Figure 4.3 A picture of an MH Acoustics Eigenmike em64, which is a sixth-order Ambisonic microphone. Inside the 84-mm diameter sphere are 64 omnidirectional microphones. (Image courtesy of MH Acoustics LLC.)

Ambisonic microphone in a fixed location is that the reproduced soundfield cannot be updated if the user moves through a space, that is, if the system provides 6DoF. In other words, a single Ambisonic microphone can offer only 3DoF in VR. However, in the next sections, we will discuss microphone arrays that can overcome this limitation.

4.3.4 *Spaced Multi-Microphone Arrays*

Multi-microphone arrays are another option to capture an overall auditory scene in a coherent way that can generate phantom image source placement without the need to pan individual sound sources or add reverberation. The main advantage of spaced multi-microphone approaches is the sense of spaciousness they can create and their ability to capture ambient or diffuse sound environments in which sound localization is less important. Multi-microphone arrays provide listeners with a more realistic impression of an acoustic space and a large sound source's size without the need to generate artificial acoustical characteristics. The main disadvantage of using multi-microphone arrays is that the resulting recordings may not offer a straightforward way to change sound source locations and listeners' perspectives as a user navigates a virtual environment, although we are not without options as we will discuss below.

4.3.4.1 ESMA

One example of a spaced multi-microphone technique is the equal-segment microphone array (ESMA) shown in Figure 4.4(a). Hyunkook Lee (2019) at the University of Huddersfield investigated the horizontal localization performance of quadraphonic near-coincident and coincident microphone configurations in loudspeaker and binaural reproductions. The work focused on Michael Williams's microphone array of four cardioid microphones arranged in a square, with a subtended angle of 90° between adjacent microphones. This array is referred to as an "equal segment microphone array" or ESMA, since the subtended angle between each pair of microphones is equal. Lee tested four different microphone spacings: 50 cm, 30 cm, 24 cm, and 0 cm. The 24 cm microphone array spacing happens to be identical to Theile's "IRT-Cross," a four-channel microphone array that originally specified a microphone spacing of 20 to 25 cm

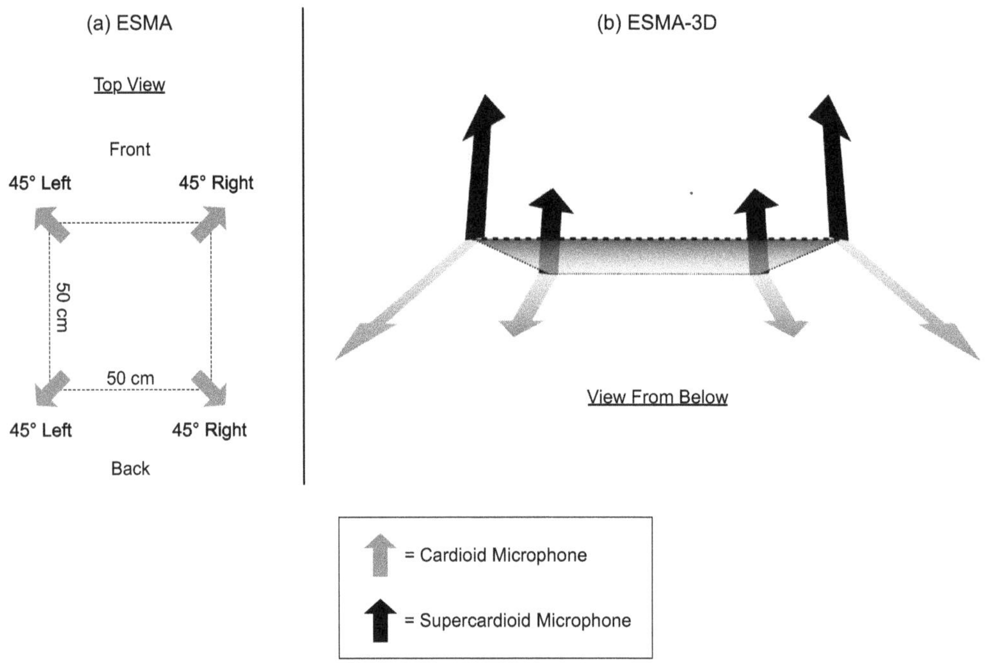

Figure 4.4 The equal segment microphone array (ESMA) for horizontal or three-dimensional capture. Diagram (a) on the left shows the top view of the original ESMA version for two-dimensional horizontal-only capture with four cardioid microphones (shown as gray arrows) arranged in a square and aimed outwards at 45°, 135°, 225°, and 315°. Diagram (b) on the right shows the ESMA-3D. The ESMA-3D technique adds four supercardioid microphones (shown as black arrows) aimed upwards and positioned coincident with the four cardioid microphones (shown as gray arrows) of the ESMA technique. The four cardioid microphones are also angled down slightly. Upward-facing cardioid or figure-8 microphones can be used as alternatives to the supercardioid microphones. Adapted from Lee (2019, 2021).

(Theile 2001). Through a series of listening tests with loudspeaker reproduction and separately with headphone reproduction, Lee found that the near-coincident or spaced microphone techniques result in considerably better localization accuracy and consistency than the coincident techniques, such as Ambisonics. Overall, loudspeaker and binaural reproduction produced similar responses and perceived source angles, but binaural reproduction resulted in more front-back confusions than loudspeaker reproduction. He hypothesized that head tracking might resolve these issues. Another benefit of near-coincident microphone techniques such as ESMA is the fact that the signals are slightly decorrelated, which can contribute to a sense of spaciousness, which can be critical for 3D audio in general but especially for ambiences and background sounds.

Lee also proposed a variation on the quadraphonic ESMA to allow vertical extension. Called the ESMA-3D, it adds four upward-facing supercardioid microphones, as shown in Figure 4.4(b). Other variations of the technique call for four upward-facing figure-8 or cardioid microphones arranged coincident with the four cardioid microphones of the ESMA technique.

Enda Bates (2018) at Trinity College Dublin introduced a variation on the ESMA technique to optimize for six degrees of freedom (6DoF) recording. Bates's technique replaces the four cardioids in the original ESMA array with four Ambisonic microphones. The technique, referred to as Perspective Control Ambisonic Microphone Array or PCAMA, utilizes dynamic steering of microphone signals from each of the four Ambisonic microphones dependent on a listener's movement through the reproduction area. The PCAMA technique can be used to control the perceived direction and distance of sound sources in response to listener movements.

4.3.4.2 Hamasaki Cube

Kimio Hamasaki and Wilfried Van Baelen (Hamasaki and Van Baelen 2015) proposed another three-dimensional multichannel microphone system. The Hamasaki Cube microphone array, as shown in Figure 4.5, consists of a lower layer and an upper layer. In the lower layer are four figure-8 microphones facing the sides, arranged in a square, and spaced 2–3 m (80–118 in.) apart. These four microphones are routed to the L, R, L surround, and R surround ear level channels. Because the array uses figure-8 microphones aimed to the sides, any direct sound reaching these microphones (assuming the direct sound originates from the front) is attenuated with the reflected and reverberant sounds from the sides accentuated. In the upper layer, four supercardioid microphones aimed upwards are arranged in a square matching the dimensions of the lower layer. These microphones are routed to the corresponding four height channels. A fifth supercardioid in the upper layer is placed in the center of the square which can be routed to the top, overhead speaker. Listening tests that Hamasaki and Van Baelen conducted indicate that listeners prefer the three-dimensional version over the original two-dimensional Hamasaki Square in terms of spatial sound quality, reality, and presence.

4.3.4.3 Spaced Ambisonic Microphones

What about a hybrid approach that combines Ambisonic microphones with spaced microphone techniques? Wieslaw Woszczyk and Paul Geluso proposed the use of spaced tetrahedral microphones to integrate direct and ambient sound and optimize the continuity of the three-dimensional immersive sound for the listener (Woszczyk and Geluso 2018).

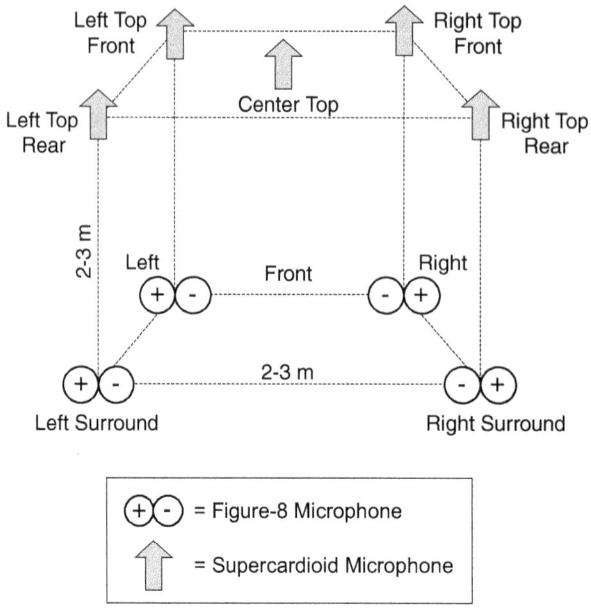

Figure 4.5 The Hamasaki Cube microphone array for three-dimensional ambience recording. The four figure-8 microphones in the lower layer are side-facing, and the five supercardioid microphones in the upper layer are upward facing. Panning locations are indicated next to each microphone. Adapted from Hamasaki and Van Baelen (2015) and Lee (2021).

Since a single Ambisonic microphone occupies a small point in space, it can be a challenge to balance sound sources and perspectives using just one microphone since some sound sources may be closer to the microphone than others. Thus, the authors proposed the use of two spaced tetrahedral microphones in an A-B (spaced left/right) configuration to capture the left and right zones of the musical ensembles. Their method is unusual since it uses raw A-Format signals directly from the four capsules, as opposed to converting the microphone outputs to B-Format signals, for the mix. The authors treat each tetrahedral microphone as a coincident four-channel microphone array and do not matrix the microphone outputs into channel-based signals. In the mixdown of the microphone signals, they pan the signals based on each microphone capsule's orientation. Recall that the four tetrahedral microphone capsules are oriented as:

- Front, Left, and Up: FLU
- Front, Right, and Down: FRD
- Back, Left, and Down: BLD
- Back, Right, and Up: BRU

In Woszczyk and Geluso's subsequent mix in a 9.1-channel system, they panned the left side tetrahedral microphone as follows:

- Back, Left, and Down (BLD) capsule: panned in between the surround left and front left channels
- Front, Right, and Down (FRD) capsule: panned between center and left channels

- Front, Left, and Up (FLU) capsule: panned to front left top channels
- Back, Right, and Up (BRU) capsule: panned to right rear top surround

The right-side tetrahedral microphone is similarly panned except that sides were reversed. The authors report that this routing produced a strong immersive effect in a 9.1-channel environment.

Perhaps related to this spaced Ambisonic microphone technique, Mróz et al. (2021) describe a volumetric recording technique in which multiple higher-order Ambisonic microphones are arranged within a space among members of an orchestra. Once the audio is captured users can navigate the space and the audio is interpolated between microphones based on tracking information about the user's location. McCormack et al. (2022) describe similar work they have done using multiple Ambisonic microphones distributed within a space to provide 6DoF capability.

4.3.4.4 ORTF-3D

Helmut Wittek and Gunther Theile (2017) developed a microphone arrangement that they say is optimal for ambient sound in 3D audio. Building on the stereo ORTF microphone technique, which specifies two cardioid microphones spaced apart 17 cm (6.7 in.) and angled 110°, the 3D technique uses four ORTF pairs for a total of eight supercardioid microphones. Two ORTF pairs of microphones face the front, and two ORTF pairs face the back. In the front-facing ORTF pairs, one ORTF pair is aimed slightly down towards the source and the other ORTF pair is angled up. There is a 90° angle between the lower and upper facing ORTF pairs. The left microphones on the front-facing pair are coincident with each other, and the right microphones on the front facing pair are coincident with each other. The rear-facing ORTF pairs are arranged similarly to the front-facing ORTF pairs, with one pair angled up and one pair angled down. In terms of panning, the downward-facing microphones are panned to L, R, L surround (LS), and R surround (RS), and the upward-facing microphones are panned to the corresponding L, R, LS, and RS height channels. In other words, the microphones in the ORTF-3D array map to eight loudspeakers positioned at the corners of a cube.

4.3.4.5 OCT-3D

The OCT 3D microphone array arose from the original Optimized Cardioid Triangle (OCT) for three-microphone stereo and the subsequent OCT-Surround (Theile 2001) technique. The OCT-3D technique adds a higher layer microphone array for height channels. The technique consists of:

- Left/Right: two supercardioid microphones, one aimed 90° left and the other aimed 90° right, panned L and R, respectively, and spaced apart 70 cm (28 in.)
- Center: a cardioid microphone is placed 8–12 cm (3-5 in.) in midway between and in front of the L and R microphones
- Surround: two rear-facing cardioids, placed 40 cm (16 in.) behind the front pair, and spaced 100 cm (40 in.), serve as the surround microphones
- Height: the upper layer consists of four upward facing supercardioid microphones 100 cm (40 in.) above the main layer

4.4 Mixing and Rendering Immersive Audio

Whether working with binaural or multichannel audio, one of the big challenges in rendering immersive audio equally well for all users is one of audio mix translation. The term "translation" in the context of audio engineering refers to the goal of making a mix sound as consistent as possible across a range of different playback systems and environments. A mix that translates well is one that sounds relatively consistent from one playback system to another. Translation has long been an issue with traditional two-channel stereo mainly when it comes to timbre, but translation of spatial attributes and sound source localization across different types of playback systems has a different set of challenges and solutions.

Let's consider an immersive audio mix created for a 9.1.4 channel monitoring system. What happens if we want to play that mix back in a monitoring environment with a different number of channels? The new listening room may have fewer reproduction channels (perhaps it is a 5.1-channel system) or it may have more channels (such as 11.1.4). Or perhaps the channel count is the same, but the loudspeakers are in different locations. Even if we can determine how to distribute the channels to the new reproduction system, we may find that the mix sounds different in the new reproduction system. Sound source locations and spatial impression of the mix in a new loudspeaker system may not be the same as they were in the original system. If either the number of loudspeakers is different, the loudspeakers are in different locations, or both, there will be a translation problem.

One solution to the translation problem is to create mixes for a variety of different playback environments, but this seems impractical because doing so takes more time and costs more money. Furthermore, it's not clear-cut which multichannel configurations engineers would choose or how they would ensure that the listener receives one that matches their playback system. To think about this issue and explore solutions, let's look at the three types of spatial or immersive audio reproduction systems:

- Channel-based audio
- Object-based audio
- Scene-based audio

4.4.1 *Channel-Based Audio*

Channel-based audio matches the number of channels of audio delivered to the listener to the number of playback loudspeakers. There is one channel of audio for each loudspeaker and there is no need for audio signal processing, mixing, or other modification at the listener's end. Stereo recordings are distributed with two channels of audio—left and right—and listeners audition the two channels using either headphones or a pair of loudspeakers. The assumption is that listeners have two channels for playback, although perhaps they listen through a single, mono loudspeaker, in which case it's best if the engineer has checked the recording for mono-compatibility, which is another translation issue. With the proliferation of single channel Bluetooth speakers, mono loudspeaker listening has no doubt had a resurgence among listeners. An older channel-based recorded media format such as a Blu-ray Disc can hold up to eight channels of audio, which typically comprises seven full-range channels and one low-frequency subwoofer channel, denoted as 7.1 channels. In channel-based formats, the mixing engineer determines the panning for each source and then renders the audio to the desired output format, whether that is stereo, 5.1, 7.1, or some other channel format. Essentially the engineer is providing loudspeaker feeds. To hear the mix as the engineer intended, the listener also needs to have the same number of loudspeakers placed in the same locations around the listening area. A playback device, such as a Blu-ray player, simply takes each channel on the disc and sends it to the corresponding audio output with the assumption of a one-to-one relationship between the number of loudspeakers in the mixing engineer's playback system and the number of loudspeakers on the listener's end. There is a further assumption that listeners' loudspeakers are placed in the same relative location as the mixing engineer's loudspeakers, which may be true in some instances, but we can't assume that for all.

To move a sound source in a channel-based system, the engineer applies a process called pair-wise amplitude panning. In stereo, this is simply the panning of the sound between the left and right channels, where the balance of a sound source is reduced in one and increased in the other to create the illusion of the sound source moving from one side to the other. The same principle applies, albeit with greater complexity, when a sound is panned in a channel-based system with a greater number of speakers. Luckily, there is panning software designed to work with common channel-based systems like 5.1 and 7.1 so that engineers don't have to deal with the complex process of pair-wise panning across more than two channels. Nevertheless, the resulting spatial movement is only compatible with the target channel count, since the process will yield channel feeds specific to that number of speakers. In other words, a separate mix would have to be made for a different channel configuration.

One approach to more flexible spatial panning within the channel-based audio paradigm is vector base amplitude panning (VBAP) (Pulkki 2001). VBAP generalizes the concept of pair-wise panning to arbitrary loudspeaker configurations by selecting the pair (for 2D layouts) or triplet (for 3D layouts) of loudspeakers whose vectors form the most appropriate base in the direction of the virtual source and computing gain coefficients for them based on vector analysis. Unlike simple left-right stereo panning, VBAP supports spatial positioning over a wide range of loudspeaker layouts, including irregular or custom arrays in two or three dimensions. While VBAP still relies solely on amplitude differences (without introducing phase shifts or time delays), it assumes accurate loudspeaker placement and listener positioning for optimal performance. As such, its spatial accuracy can diminish if the listener is not located at the sweet spot or if the loudspeaker configuration deviates from the intended setup.

4.4.2 *Object-Based Audio*

The next category of sound reproduction systems is object-based audio, which has advantages over channel-based audio for immersive applications. In contrast to channel-based audio, which requires engineers to deliver separate mixes for each playback system, object-based audio allows any mix to be played back on any loudspeaker layout. Audio "objects" are simply audio signals (or tracks), such as prerecorded tracks from spot microphones, sound synthesis engines, Foley, or any type of sound-designed track, with associated metadata that determines how each audio track should be rendered during playback. The metadata specifies sound source location and any associated processing such as equalization, compression, and reverberation, any of which may vary in time or with user interaction. The ITU-R has standardized the structure of the metadata describing the technical content and format of audio in the Audio Definition Model (ADM) in Recommendation ITU-R BS.2076 and has further specified audio renderers for ADM in Recommendation ITU-R BS.2127 (International Telecommunication Union 2019, 2023).

We can think of audio objects as virtual loudspeakers located around the listening position. The main benefit of object-based audio is that it doesn't matter that mixing engineers and listeners have different playback systems because the listeners' playback systems receive instructions (in the form of ADM metadata) to pan and render tracks to the locations specified by the mix engineer. An engineer delivers a mix that includes not only audio tracks but also metadata that instructs a listener's sound system to pan sounds determined by the mixing engineer using information about the number of loudspeakers and loudspeaker placement in the listener's system. For example, if a mix includes the sound of a car that moves from left to right, the rendered audio will include not only the car audio but also instructions for the playback system to recreate the panning that the mixing engineer specified over whatever loudspeaker array is available to the listener. The listener needs to specify in their playback system processor how many loudspeakers are present and where they are located. With this information, the listener's playback system processes the sound appropriately to achieve the mixing engineer's intent or at least a close approximation. Object-based audio does not arrive completely mixed, rather the final mixing happens in the listener's compatible processor just prior to the audio being sent to the listener's loudspeakers.

The main disadvantage of object-based audio is that each audio object or channel to be rendered at the listener's end requires at least one channel on the recording medium or transmission path. With channel-based audio, a fixed number of channels (stereo, 5.1, 7.1.4, etc.) are transmitted so at least we can anticipate the bandwidth required for transmission, even though the audio channels may not align with the listeners' specific playback systems. Object-based audio, by contrast, may require hundreds of channels to be transmitted and rendered at the listeners' end. A potentially unlimited number of audio objects is not really a feasible option because of finite transmission bandwidth and user-end processor power.

Several object-based audio formats exist, including AuroMax, Dolby Atmos, DTS:X, and MPEG-H, although Dolby Atmos uses a combination of objects and what are known as bed tracks. Bed tracks are channel-based audio and can range in number from 2.0 (two-channels) to 7.1.2 (seven floor channels, two overhead channels, and one LFE channel).

4.4.3 Scene-Based Audio

Scene-based audio, as described in the European Broadcasting Union (EBU) Technical Review (Olivieri et al. 2019), offers an elegant solution to the shortcomings found in channel-based and object-based audio formats. Scene-based audio combines the advantages of channel-based audio—a limited and fixed number of channels—with object-based audio, which means listeners' playback systems do not need to match the mixing engineer's playback system. Like object-based audio, scene-based audio works with various loudspeaker configurations and is rendered on the listener's end. Ambisonics is an example of scene-based audio in which a channel represents a component of the soundfield rather than a specific loudspeaker. As we discussed earlier, first-order Ambisonics has only four channels of audio, but it can be played back on any number of loudspeakers. Similarly, a higher-order Ambisonic signal can be decoded to fewer number of speakers than the channels available in the B-format Ambisonic signal. Decoding an Ambisonic multichannel signal is based on the playback system speaker layout. In the next section, we will discuss encoding and decoding scene-based audio mixes via Ambisonics.

4.4.4 Encoding and Decoding Ambisonics

Traditionally, surround sound is mixed to a set number of output channels, corresponding to the number of channels available in an audio playback system. As we discussed above, this is known as channel-based audio, where each type of playback system (e.g., stereo, 5.1, 7.1, 9.1, etc.) requires a different mix according to the number and placement of loudspeakers. Ambisonics, on the other hand, allows us to decouple the number of output channels of the mix from the number of channels in a playback system. In other words, we can create one Ambisonic mix of any order and decode it for playback on almost any multichannel audio system or binaurally over headphones without the need to remix it. A tenth-order Ambisonic mix with hundred and twenty-one channels of audio can be played back on any stereo or surround sound system as can a first-order Ambisonic mix with four channels. This is because spatial information (e.g., panning and elevation) is encoded in the B-Format multichannel signal no matter the Ambisonic order.

Let's look at the basic signal flow building blocks needed for Ambisonic encoding and decoding. Figure 4.6 is a block diagram showing the basic, abstracted signal flow from input encoding to output decoding. The diagram shows two types of input sources: mono/stereo signals (in the top left) and Ambisonic microphone signals (in the bottom left). Input signals are converted to B-Format Ambisonics which is an intermediary soundfield representation. As we mentioned earlier, this soundfield representation, regardless of its order, can be decoded to any loudspeaker layout or binaurally to headphones. Likewise, the intermediary soundfield representation can be converted to a different Ambisonic order. Up-converting a mix, say from first-order Ambisonics to third-order Ambisonics, does not provide any additional spatial resolution in a mix that has already been encoded in first-order Ambisonics. Any blurriness in sound source position remains and the higher-order version has the additional downside of more channels. For example, a mix in first-order Ambisonics with four channels becomes a 16-channel mix when converted to third-order Ambisonics. Down-converting a mix, say from seventh-order to third-order Ambisonics, reduces the spatial resolution somewhat but still may be acceptable for a project's needs. A down-conversion has the benefit of a lower channel count (64 channels for seventh-order to 16 channels for third-order), which saves data transmission and storage requirements as well as processing power. If reducing the Ambisonic order is necessary, it is best to listen to a project at the lower order to determine what impact any reduction in spatial resolution may have on it.

To mix a project in Ambisonics, a digital audio workstation (DAW) with multichannel track capability, such as Avid Pro Tools (Ultimate or Studio), Cockos REAPER, or Merging Technologies Pyramix, or a virtual patching environment such as Cycling '74 Max (also known as Max/MSP) is required. REAPER seems to be the DAW of choice for many Ambisonics researchers, sound designers, and recording engineers, perhaps due to

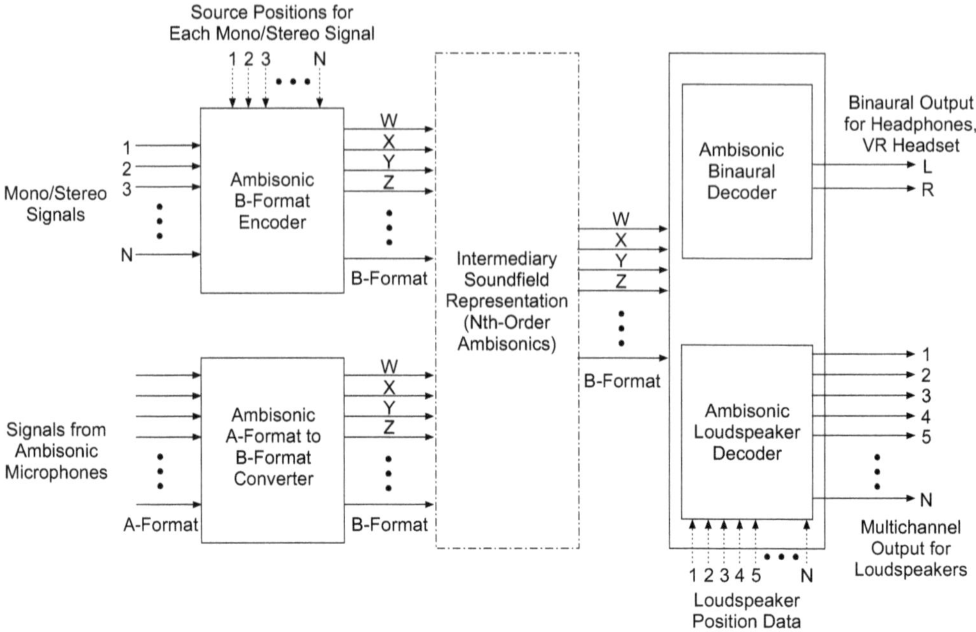

Figure 4.6 A block diagram showing an abstracted Ambisonic signal flow from input encoding to output decoding in a digital audio workstation. Input sources, either mono/stereo signals or from Ambisonic microphones, are first converted to B-Format Ambisonics (shown on the left side of the diagram). B-Format is an intermediary soundfield representation (shown in the middle of the diagram) in any Ambisonic order number. The intermediary soundfield representation can then be decoded to any loudspeaker layout (and any number of loudspeakers) or binaurally to headphones (shown on the right side of the diagram).

its capability to work with high channel count tracks and its relatively low cost. The post-production stages of our case studies, described in Sections 5.5.4 and 5.6.5, utilize REAPER to showcase Ambisonic processing and mixing. REAPER version 7 and higher allows up to 128 channels per track, which supports up to tenth-order Ambisonics if the DAW host computer has sufficient processing power to handle 121 channels on each track and mix bus. The number of channels in each track and bus needs to be set to match the Ambisonic order with which we are working. Recall that first-order Ambisonics requires at least 4 channels, second-order requires at least 9 channels, third-order requires at least 16 channels of audio, and so on. The relatively high number of channels per track required for Ambisonic projects contrasts with conventional stereo projects in which DAW input/mix tracks are typically either mono (one channel per track) or stereo (two channels per track) and the mix output bus is stereo.

Within a DAW, Ambisonic plugins are required both at the front end of the signal flow, usually in the form of an Ambisonic panner (for mono or stereo signals) or encoder (for A-format Ambisonic recordings) on each mix or input channel, as well as on the monitoring or output mix bus, usually in the form of a multichannel decoder for loudspeakers or binaural decoder for headphones. To have the capability to pan every mix track to any point in the Ambisonic spherical soundfield, each track needs to have an Ambisonic panner/encoder. The Ambisonic order of a panner/encoder's order should ideally match the input order of the decoder, although a lower-order Ambisonic signal could be fed into a higher-order decoder as, for instance, the four channels of a first-order Ambisonic signal would constitute the first four channels of a second-order signal. We will further discuss these order relationships in practical terms in Section 5.6. Finally, the output order of the decoder should match the Ambisonic order required for project delivery.

Additional signal processing plugins such as equalization, compression/limiting, and reverberation can also be inserted on each mix/input track. A mix bus between the mix/input channels and monitoring busses allows for overall mix processing, such as with Ambisonic compression, EQ, and reverberation plugins, as well as routing to multiple decoding outputs. Plugins on the Ambisonic mix bus should have the capability to process the number of Ambisonic channels required by the Ambisonic order of the project. For example, a mix bus on a seventh-order Ambisonic project requires plugins that support at least 64 channels or seventh-order Ambisonics. Figure 4.7 is an abstracted block diagram showing signal flow within a DAW project.

In addition to maintaining matched signal levels for all channels within an Ambisonic signal or track, we also need to ensure that any processing such as equalization, compression, or delay is applied equally across all channels within an Ambisonic signal. Even a sub-millisecond delay in one channel of an Ambisonic signal can cause the spatial relationship between channels to change and therefore alter the intended location of sources. Anytime we have mono or stereo sources to be mixed into an Ambisonic soundfield, we can apply processing to these individual mono or stereo tracks prior to the Ambisonic encoding/panning stage, which can be less processing-intensive than applying the same processes to the multichannel B-format signal.

Several Ambisonic encoding/panning plugins and decoding plugins are available, including the following, listed with the highest order of Ambisonics they support, the channel order convention used, and the supported plugin format:

- Blue Ripple Sound O3A Core plugin library (free)
  - includes encoder/panner and decoder plugins
  - up to third-order Ambisonics (16 audio channels) using the AmbiX convention
  - VST2/AAX
- Blue Ripple Sound O7A Core plugin library (paid)
  - includes encoder/panner and decoder (including for binaural) plugins
  - up to seventh-order Ambisonics (64 audio channels) using the AmbiX convention
  - VST2/AAX
- DearVR PRO 2 plugin (free)
  - combined encoder/panner, decoder (including for binaural), reverb, and reflections plugin
  - up to third-order Ambisonics (16 channels) with both AmbiX and FuMa output formats
  - VST3/AU/AAX
- ICST AmbiEncoder and ICST AmbiDecoder plugins (free)
  - open-source Ambisonic encoder/panner and decoder plugins developed at the Institute for Computer Music and Sound Technology in Zurich, Switzerland
  - up to seventh-order Ambisonics (64 audio channels) using the AmbiX convention

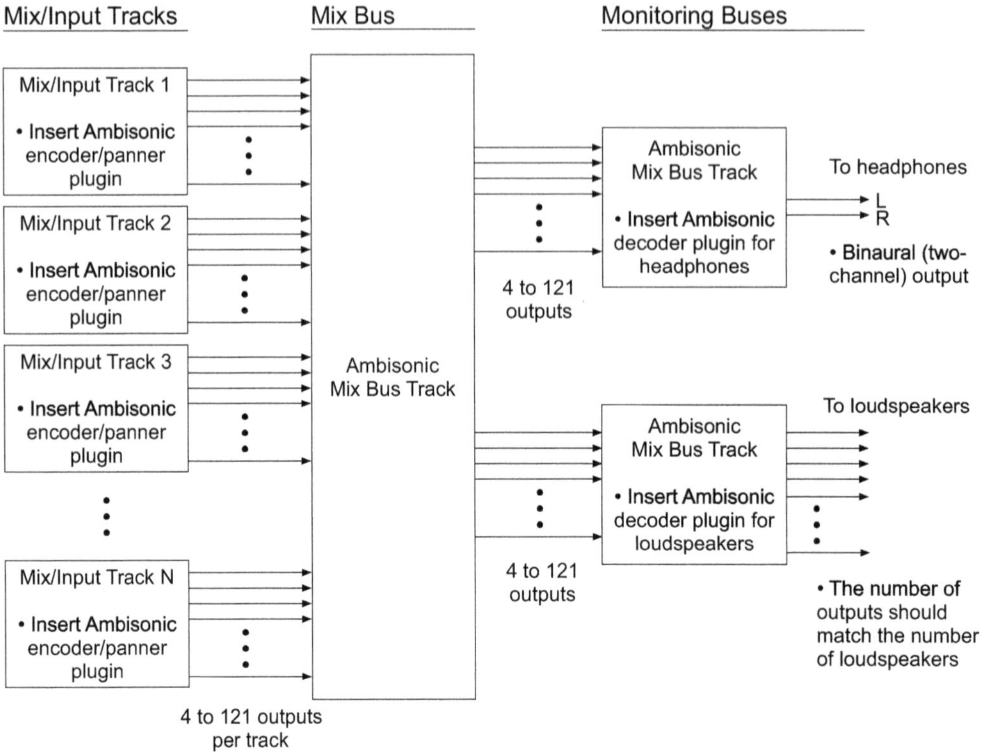

Figure 4.7 A block diagram showing Ambisonic signal routing within a digital audio workstation. Once the Ambisonic order number is determined for a project, each plugin should be set to match the Ambisonic order number of the project. It is also crucial to set each mix/input track (shown on the left side of the diagram) and each bus track (shown in the middle and the right side of the diagram) to the required number of channels per track. The number of channels per track is determined by the Ambisonic order number. First-order Ambisonics has four channels, second-order has nine channels, third-order has 16 channels, and so on up to tenth-order which has 121 channels. Besides the required Ambisonic encoder and decoder plugins at the beginning and end of the signal chain, optional plugins for equalization, compression/limiting, and reverberation can be inserted on mix/input tracks, which are likely to be mono or stereo audio sources, and on the Ambisonic mix bus track. For plugins inserted in the Ambisonic mix bus, set the number of plugin channels to match the number of channels on the track (as determined by the Ambisonic order).

  - The ICST AmbiDecoder plugin includes output options for several different loudspeaker configurations, but it does not include a binaural renderer.
  - VST3/LV2
- IEM Plug-in Suite plugin library (free)
  - includes encoder/panner and decoder (including for binaural) plugins (IEM StereoEncoder and IEM MultiEncoder), as well as a combined encoder and room simulator plugin (IEM RoomEncoder)
  - open-source plugins for 3D audio and Ambisonics developed at the Institute of Electronic Music and Acoustics in Graz, Austria

    - up to seventh-order Ambisonics (64 audio channels) using the AmbiX convention
    - VST2/LV2
- SPARTA Collection plugin library (free)
    - includes encoder/panner and decoder (including for binaural) plugins
    - open-source Ambisonic plugins developed at Aalto University, Finland
    - up to tenth-order Ambisonics (121 channels) using the AmbiX convention
    - VST/LV2/VST3
- SSA a1Panner (free), a3Panner (paid), and aXPanner (paid)
    - SSA a1Panner: up to first-order Ambisonics (4 channels) using the AmbiX convention
    - SSA a3Panner: up to third-order Ambisonics (16 channels) using the AmbiX convention
    - SSA aXPanner: up to tenth-order Ambisonics (121 channels) using the AmbiX convention
    - AAX/VST3
- Waves B360 Ambisonics Encoder plugin (paid)
    - can convert from mono, stereo, or surround tracks to first-order Ambisonic B-Format
    - up to first-order Ambisonics (4 channels)
    - includes AmbiX-to-FuMa and FuMa-to-AmbiX conversion utilities
    - AAX/VST3

The encoder/panner is a critical Ambisonic plugin type since it allows us to encode and pan sources to any location on the surface of the virtual spherical soundfield around the listening location, at whatever Ambisonic order the plugin supports. As far as we are aware, the highest order of any plugin is tenth-order Ambisonics, which is supported by the SSA and SPARTA plugins. An encoder/panner plugin also allows us to convert a channel-based surround mix (e.g., 5.1, 7.1, 9.1.4) to an Ambisonic B-Format by placing the plugin on a multichannel mix track or surround mix bus and panning the channels to the standard loudspeaker locations. A screenshot of the IEM StereoEncoder plugin is shown below in Figure 4.8 with a source panned to the upper left quadrant. We will discuss processing spot microphones using an IEM encoder plugin further in Section 5.6.5.3.

4.4.5 Ambisonic Reverberation

Room simulation or virtual acoustics are important for creating a sense of distance and auditory perspective in a virtual scene, and this is often accomplished with reverberation plugins. Most such plugins are either algorithmic or convolution-based and can be computationally intensive due to the complex signal processing involved. As such, it is helpful to use them in the most efficient way possible. In traditional stereo mixing, instead of placing the same reverb plugin on each track, the tracks are routed to an auxiliary track, in which a single instance of the plugin is inserted, resulting in a more efficient use of processing power. With an Ambisonic mix for VR, the same approach can be followed, wherein a multichannel Ambisonic reverb plugin can be applied to the B-format signal.

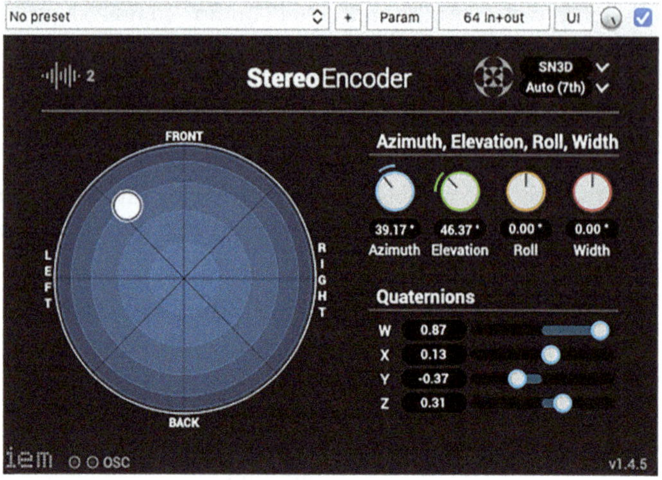

Figure 4.8 A screenshot of the IEM StereoEncoder for panning/encoding mono or stereo audio signals into Ambisonics and supports up to seventh-order encoding. The plugin indicates the order number ("Auto 7th") in the upper right corner. Sound source location can be controlled using the circular panner on the left (which shows a source panned to the upper left quadrant), by adjusting azimuth, elevation, roll, and width values (all expressed in degrees), or by adjusting the B-Format quaternions (W, X, Y, and Z). When any one of these controls is changed, the others are updated to reflect any change in source location.

However, when moving sound sources are encoded into Ambisonics, it may be helpful to have a reverberation plugin on each track so that reverberation parameters can be updated to reflect the changing locations of each sound source. While this approach is processor-intensive, it allows the simulation of virtual acoustics that respond to sound source location. This is why plugins such as dearVR PRO 2 and IEM RoomEncoder, shown in Figure 4.9, incorporate encoding with reflections and reverb in the same plugin. Early reflection patterns that update with sound source location can reinforce perceived sound locations within a virtual room and help listeners estimate the size of the room they are in.

4.4.6 Binaural Rendering

To monitor Ambisonic audio requires either a multichannel decoder to listen back through a loudspeaker system, or a binaural decoder to listen over headphones. Decoders need to be set to the correct Ambisonic order, channel order format (FuMa or AmbiX), and normalization setting, the most common of which is SN3D.

Monitoring spatial audio over headphones requires the decoder to convert Ambisonic or object-based audio channels to binaural audio. Binaural renderers typically use one of two methods to convert a multichannel mix to two-channel playback over headphones. In one method, binaural renderers encode each audio object directly with HRTF processing. Also known as direct virtualization, this process takes each panned source and renders it using HRTFs that match each source's intended location. Direct virtualization is

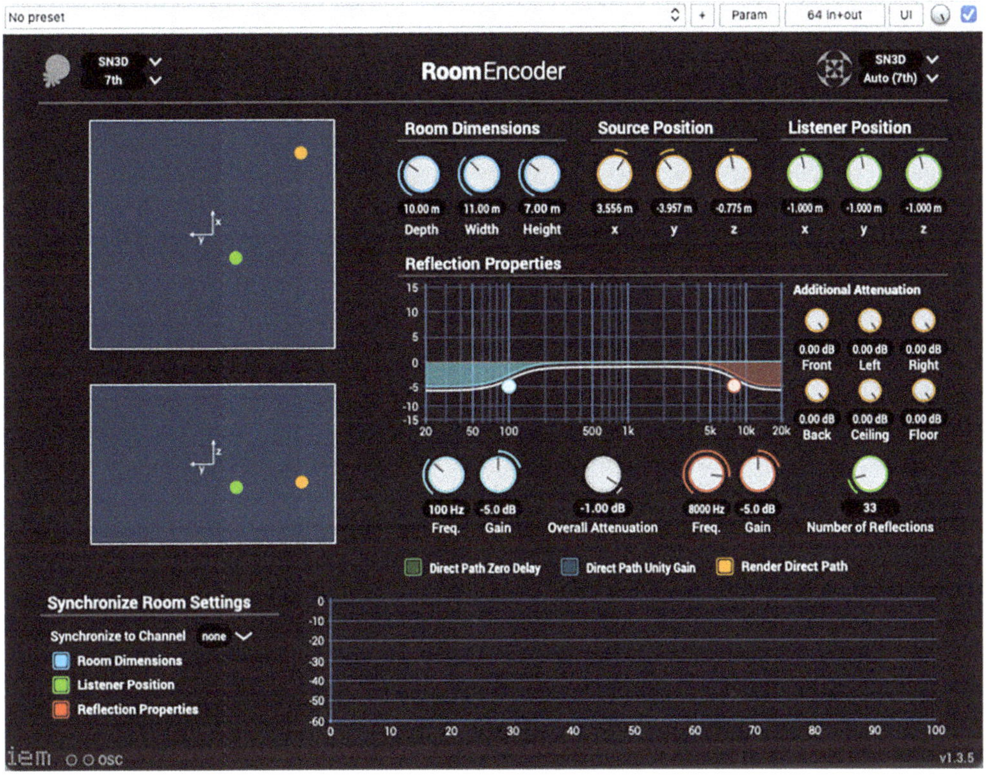

Figure 4.9 A screenshot of the IEM RoomEncoder for panning/encoding signals with room simulation for up to seventh-order Ambisonics, shown with a source panned to the front right. This plugin also allows the listener position to be changed without changing the source location. Source and listener positions can be controlled using x, y, and z values at the top or by using the graphical panners on the left. Room reflections are updated as the source or listener change position.

the most accurate way to render a multichannel mix, but it can be limited by the user's VR system since processing requirements increase proportionally to the number of audio objects.

The other way to go from a multichannel mix to binaural audio involves a two-stage process. In the first stage, a renderer converts the panned audio objects in a mix to an Ambisonic signal, which is then converted to a binaural signal by a binaural decoder. One advantage of this method is that by converting to Ambisonics first, it reduces the number of possible channels or sources to be binaurally rendered. For example, if a mix has been converted to a first-order Ambisonic signal, there are only four channels to render binaurally, compared with perhaps hundreds of audio objects in the original mix that would all need to be individually rendered with HRTFs if using the direct virtualization method. While this two-stage method saves computational power, the tradeoff is that we may lose spatial resolution. We need to be especially concerned about losing spatial resolution if we convert to first-order Ambisonics first, since first-order has only four channels of audio and limited spatial resolution. In other words, the locations of sound sources may not be as precise or accurate after converting to first-order Ambisonics as they are with direct virtualization. A compromise between these two methods is to convert to a higher-order Ambisonic signal (rather than first-order) prior to binaural rendering since higher-order Ambisonics offers better spatial resolution than first-order.

There are many binaural audio renderers available, but they are not all equal in performance. Researchers at New York University conducted listening tests on six binaural renderers and found significant differences in their localization accuracy. Side localization was problematic for most renderers except for one that uses direct virtualization and another one that is higher-order Ambisonics. Perhaps not surprisingly, they also found that first-order Ambisonic renderers performed the worst on localization accuracy overall when compared to higher-order Ambisonic renders and direct virtualization (Reardon, Genovese, et al. 2018a). The results from their listening tests on sound externalization, and front-back and up-down confusions, indicated that first-order Ambisonic renderers performed the worst, while a direct virtualization renderer and one of the higher-order Ambisonic renderers performed the best (Reardon, Zalles, et al. 2018). The research group conducted further listening tests on the same group of binaural renderers to evaluate their spatial sound quality along several dimensions, including timbral balance, clarity, naturalness, spaciousness, and dialogue intelligibility, using music recordings and movies as program material. They found that binaural renderer performance was highly dependent on content and that there was no one binaural renderer that was optimal for all sound quality attributes across the settings they tested (Reardon, Genovese, et al. 2018b). Given the variability of binaural decoders and the fact that these articles do not list plugin makes and models, it seems prudent to listen to a few decoders and choose the one that sounds the best for your project.

There are several Ambisonic-to-binaural decoders available including:

- Blue Ripple Sound O7A Decoder – Headphones, up to seventh order (paid)
- dearVR AMBI MICRO, up to third order (free)
- dearVR PRO 2, up to third order (free)
- IEM BinauralDecoder, up to seventh order (free)
- Noisemakers Ambi Head HD, up to seventh order (paid)
- Penteo 360 and Pro+, up to seventh order (paid)
- SPARTA CroPaC Binaural, first order (free)
- SSA a1Monitor, a3Monitor, and aXMonitor, up to tenth order (paid)
- Waves NX Ambisonics, first order (paid)

Some binaural decoders, such as the IEM BinauralDecoder, provide headphone equalization for a select list of headphone models. If your headphone model is on the list, selecting it should give you a more accurate headphone response, and therefore a less colored sound.

It is important to check the output format setting of the panner plugin. For example, the dearVR PRO 2 has output settings for headphones (binaural, 2-channel), Ambisonics (first-, second-, and third-order in FuMa and AmbiX formats), and loudspeakers (over 30 multichannel formats from 2.0 stereo to 13.1). If you have set up your DAW with a binaural decoder on an Ambisonic mix bus, you'll want to make sure each mix channel's encoder plugin output is set to the appropriate order of Ambisonics and not to output a binaural signal.

4.4.7 *Spatial Audio Codecs*

Tying these various spatial audio formats together are two audio codec standards: MPEG-H and MPEG-I. Developed by the Motion Picture Experts Group and Fraunhofer IIS, MPEG-H is described as a "standard for universal spatial/3D audio coding" that can transport channel-based audio, object-based audio, and higher-order Ambisonic input (Herre et al. 2014). It supports sound reproduction across a range of output formats including stereo, binaural, and 5.1, as well as audio formats with multiple layers such as 5.1.4, 7.1.4, 22.2, and beyond. MPEG-H contains functionality that allows listeners to interact with audio, for example, to adjust the position or level of mix elements, such as a dialog or an announcer, independent of other audio elements. MPEG-H powers Sony's 360 Reality Audio and has been adopted as a broadcasting standard by South Korean TV and Brazilian TV.

Building on the MPEG-H standard, a forthcoming standard, MPEG-I, is currently under development within the MPEG Audio group (Herre and Disch 2023; Herre et al. 2024). The standard aims to provide virtual and augmented reality applications with support for 6DoF rendering, object-based audio, channel-based audio, scene-based audio higher-order Ambisonics, as well as sophisticated modeling of acoustic source characteristics including spatial extent, directivity, and Doppler effect. Synthetic room simulation effects including reverberation, early reflections, occlusion, and diffraction are also part of the standard. MPEG-I can render to not only binaural output for headphone listening but also for a wide range of loudspeaker setups for a single listener whose position is tracked. Location tracking allows the renderer to adapt the level and delay for each loudspeaker to compensate for listener position while providing a stable sound image as though the listener is always in the sweet spot of the loudspeaker array.

In this chapter, we gave an overview of sound reproduction formats, microphone techniques, and mixing and rendering methods for spatial audio. In the next chapter, we will discuss practical and aesthetic considerations in producing sound for VR film and present two case studies.

Chapter 5

Sound in VR Film

Crafting Audio for Fixed-Media VR

One of the primary applications of modern VR is in film production. An increasing number of filmmakers are utilizing 360° recording tools and devising new methods of storytelling that leverage the unique affordances of immersive media. While sound has always been a fundamental element of traditional cinema, VR film—also called Cinematic VR (Dooley 2021)— places even greater emphasis on it due to the inherently immersive nature of acoustic phenomena. In VR, sound plays a crucial role in guiding the viewer's attention in 360° space, articulating the dimensions of the VR environment, and bridging the visual and kinesthetic qualities of room-scale experiences. Just as capturing VR film introduces new technical and aesthetic considerations, sound production for VR presents its own set of challenges and creative possibilities. In this chapter, we connect our previous discussions on psychoacoustics and spatial audio to practical applications of sound recording and production for VR film. We cover the fundamental elements of film sound production, including crew roles, techniques, and equipment, with a particular focus on narrativity and immersion in VR. We then present two case studies where we prepare, capture, process, and finalize audio for VR film, highlighting how these processes align with or differ from traditional practices in film sound. These case studies illustrate the unique aspects of VR filmmaking and serve as a guide for working with 3D 360° fixed-media content.

5.1 Elements of Film Sound

Film sound relies on a wide range of sound sources and audio processing techniques. Imagine the rich assortment of sounds that you might encounter while watching a movie. You hear a conversation between two characters walking down a busy street backgrounded by the constant hum of traffic with car horns going off every now and then. Despite the cacophony, you still make out the footsteps of the characters and maybe even those of other people passing by. All of these sounds are framed by a piece of orchestral music, imperceptible to the characters in the scene yet decisive in your emotional response to the story. Each of these sounds was likely produced at different times, in different places, by different professionals, using different techniques.

Among the most foregrounded elements of film sound are spoken words and the **dialogue** between characters. Due to the critical role of these elements in conveying the story of a film, they need to be captured and processed with intelligibility in mind. Dialogue can be captured on set with a boom mic or lavalier mics that are strategically hidden in clothing, hair, or makeup. Dialogue can also be added during post-production through

DOI: 10.4324/9781003048022-5

voiceover, where the actors themselves or separate voice actors record the lines on top of the film in a studio using a process called automated dialogue replacement (ADR).

Besides dialogue, film sound involves **sound effects** and environmental sounds that emanate from the objects or events in the scene. For example, these can be engine noises, explosions, and weather-related sounds, such as rain and thunder. These could also include sounds that stem from the interaction of a character with an object in the scene, such as the sound of a door closing or a faucet being turned on. When captured on set, these sounds are referred to as production sound effects (PFX). However, not every possible sound source in a scene can be captured on set. In such cases, sounds are added to a scene later in post-production. The added sounds can be captured from real-life objects (e.g., a scene where a character operates a drill can be paired with a sound of this tool recorded separately), or they can be generated synthetically, especially when there isn't a real-life analogue of the object in a scene (e.g., the sound of a spaceship taking off in a sci-fi movie can be designed using a synthesizer).

Another method for adding sound effects to a scene is **Foley**. This technique involves recording objects that produce the same or similar sounds as those depicted on screen. For instance, instead of recording the hooves of a running horse, which can be a challenging task, the Foley artist can replicate this sound by recording two coconut halves being tapped together.[1] Foley is recorded in specialized studios where the artist will have a wide assortment of objects that can be used to recreate sound events. A defining quality of Foley effects is that they are performed on top of the film to match the visual progression of the events in a scene (Ament 2014). For example, for a scene where a character is walking on wood flooring, the Foley artist might record themselves stepping on a piece of wood as they closely follow the gait of the character.

Another significant yet understated element of film sound is **ambience**. Ambience can include room sound (i.e., the sound of a room in silence), nature sounds such as wind and rustling trees, traffic noises, machinery sounds, and the murmur of a crowd in the background (also known as walla). The ambience track can be considered as an auditory analogue of the "establishing shot" in film that provides the viewer with a context as to where a particular part of the story is taking place. The ambient sounds often help evoke a sense of place in the scene and glue together various other sound sources, giving them a shared context. As such, ambient sounds contribute to how convincing a scene feels. It can also be utilized to make a scene feel more open or more intimate. For instance, a scene by the beach could use sounds of seagulls or waves crashing. However, turning down these elements in the mix, and emphasizing the dialogue can make the scene feel more introspective. While ambience can be captured at the location of a shoot, it can also be recorded in environments similar to that location or artificially crafted with Foley and sound effects.

Last but not least, a prominent element of film sound is **music**, also referred to as the score. The score can set or heighten the mood of a scene (e.g., up-tempo music that accompanies a chase scene), convey a meaning that's not apparent on screen (e.g., mysterious music that foreshadows the danger that a character is about to face), or establish recurring themes that signify certain characters, places, or events (e.g., "The Imperial March" that plays in scenes that feature Darth Vader in *Star Wars*).

Most of these sound elements play an equally important role in VR film as they do in traditional cinema. However, the recording of these sounds and their integration into immersive media requires different tools and techniques, which we will explore in our case study later in Section 5.5. While dialogue, sound effects, and music maintain their narrative functions in VR film, ambience can play a particularly significant role in establishing a sense of plausibility in VR. Since the viewer is situated within the space where the film is set, the acoustic properties of this virtual environment will impact how the viewer relates themselves to it. When ambient elements such as room tone and reverberation, whether recorded during production or added later, closely match the visual properties of the environment, they contribute to a more convincing sense of place. This in turn provides a sense of "being there," which the VR researcher Mel Slater characterizes as the place illusion (2009). This can make the viewer feel more present in the environment shown on film. If, however, the ambience does not match the visual space presented in VR, it might instill an uncanny feeling, which can break the place illusion but could also serve as a narrative device.

The prominent use of headphones in VR systems can lead to the in-head localization effect (also known as lateralization) that we discussed in Section 3.2. This can prompt a more acute listening experience that draws the viewer's attention to subtle details in ambience. Even the quietest environments in real life will have some degree of ambient noise to them, whether it is the distant hum of appliances or the faint rustle of wind. Although we may not notice these noises until we actively pay attention to them, they nevertheless contribute to how we perceive our surroundings. This is why it is advisable to avoid total silence in VR unless it serves a narrative function. Especially with closed-back headphones that isolate outside noises, a lack of ambience can create an awkward or unsettling experience for the listener. Background noise, as quiet as it may be, can make a VR experience feel more natural.

5.2 Production Sound

Making a VR film involves a multitude of techniques informed by traditional sound and film production, on the one hand, and new creative possibilities ushered in by VR technology, on the other. Film sound practices span both the production and post-production phases of a film project. Whereas the production phase involves capturing dialogue and sounds on location, the post-production phase entails the mixing of the sounds recorded during production as well as the addition of new sounds, such as dialogue, sound effects, Foley, and music. In this section, we will concentrate on the production phase. We will first look at the makeup of a traditional film sound crew and describe the various duties undertaken by its members, followed by a discussion of how these duties translate to VR film production. We will then explore practical considerations for capturing sound for VR film.

5.2.1 Crew

A traditional film sound crew consists of three primary roles: a production sound mixer, a boom operator, and a second assistant. The **production sound mixer** leads the film sound crew, ensuring that the on-set sound sources essential for post-production are captured with the appropriate equipment during filming. In this capacity, the production sound mixer manages the balance of each audio channel and pays attention to any unwanted noises or distortions.

The **boom operator**, also known as the first assistant sound, is responsible for operating a shotgun microphone[2] on a boom pole during filming. This person maintains the correct placement and orientation of the microphone while making sure that it remains above the frame line (i.e., outside the camera's field of view). The boom operator captures a range of sounds from the dialogue between the actors to noises emanating from the events in the scene. Furthermore, they assist the production sound mixer in preparing the set for audio recording prior to each shot.

The **second assistant sound** helps the boom operator and the production sound mixer with the arrangement and maintenance of audio equipment before, during, and after a shoot. This person serves a dynamic role as they work with various other departments on set to ensure the smooth functioning of the film sound crew.

While VR film production maintains most of these duties, boom miking is often substituted with spot miking (e.g., with lapel mics attached to actors) since the boom operator cannot remain within the panoramic field of view of the VR camera. This is a more general concern in VR film production affecting many other facets of traditional film making. The "behind the scenes" of a traditional film production is often crowded with crew members and their equipment. With cameras and microphones that can capture in all directions, 360° VR film is devoid of a so-called behind the scenes. This requires creative solutions to hide the crew and equipment from the audience's field of view. With today's wireless technology, most capture devices can be operated remotely, eliminating the need for crew members to helm their equipment, as long as the equipment itself can be hidden from the camera.

In location shooting, the natural or architectural features of an environment can be leveraged to hide the crew. In areas where such features aren't available, the crew can be stationed behind green-screened zones that can be edited out of the scene in post-production. Shoots on a film studio will offer more flexibility with devising a set where the crew and equipment can be hidden. Nevertheless, it is still a technical and creative challenge to set up a fully panoramic set and strategize where to place the capture devices. Additionally, unlike location shooting where natural light might be sufficient for filming, a film studio will require artificial lighting, which will inevitably be visible to a 360° camera and will need to be removed from the scene in post-production. In Section 5.5, we will offer one such filming scenario as a case study, discussing the workarounds that can be utilized to deal with the limitations of VR filming.

5.2.2 *Capturing a Soundfield*

In Chapter 4, we discussed soundfield techniques in spatial audio with a particular emphasis on Ambisonics. You may remember that Ambisonics can be used for encoding sounds into a 3D soundfield and decoding the result to various output systems from headphones to speaker arrays. This makes Ambisonics a powerful intermediary format for immersive audio delivery. We also discussed that Ambisonics can also be used for capturing sounds with specialty microphones called Ambisonic (or SoundField) mics. An Ambisonic mic captures its surroundings in 3D, allowing the listener to rotate their head within the captured soundfield much like how we rotate our heads in VR to view different areas of the virtual space. This is why Ambisonic recording can be a good companion for VR film, especially for capturing an overall auditory representation of a scene.

As we discussed in Section 4.3.3, a first-order Ambisonic microphone offers a tetrahedral combination of four condenser capsules. While this configuration ensures a uniform pick-up pattern across all three dimensions of the recording environment, the body of the microphone and any mounting mechanism can still cause occlusions, which need to be considered while positioning the microphone. Depending on the acoustic space and the placement of the sound sources in it, an Ambisonic microphone can be positioned vertically or in an end-fire (horizontal) configuration. In the vertical placement scenario, the microphone can be set up in upright or upside-down orientations. When possible, it is advisable to position the Ambisonic microphone in an upright position to mirror the orientation of the human auditory system, wherein the occlusion caused by the microphone's body resembles that of the human torso. This orientation also offers the benefit of rejecting floor reflections to some degree. The end-fire placement, where the microphone is parallel to the ground, can be useful for 180° VR film. In this scenario, the microphone and camera would face the same direction, allowing the body occlusion of the microphone to be on the same side as the area behind the camera's field of view.

5.2.3 *Capturing Point Sources*

In Chapter 4, we discussed various spatial audio recording techniques, including those that involve Ambisonic mics. While their ability to capture surround sound makes these techniques appropriate for VR production, none of them can offer a catch-all solution for recording sounds in VR film. For instance, a first-order Ambisonic mic, such as the Sennheiser Ambeo or the RØDE NTSF-1 SoundField microphone, can be great for capturing an immersive ambience. However, the four-channel spherical decomposition of a soundfield captured with a first-order Ambisonic mic may not provide enough spatial resolution to accurately localize point sources.[3] Prior research shows that while a first-order Ambisonic reproduction can perform reasonably well with frontal sound sources, lateral sources suffer from greater spatial ambiguity and offer less localization accuracy than higher-order Ambisonics (Bertet et al. 2013).

The three-dimensional ambience captured with a first-order Ambisonic microphone can be augmented with the monophonic feed from spot microphones that capture dialogue or other sound sources that need to be clearly intelligible. As we touched upon earlier, since a VR camera captures a panoramic view of the world around it, there is no frame line above which a boom mic can be placed to record point sources. Instead, spot recordings are made with lavalier microphones that can be hidden under clothing or makeup. These recordings can then be encoded into a higher-order Ambisonic soundfield for greater localization accuracy than what a first-order Ambisonic mic alone could provide. Moreover, since the spot microphones are bound to be more closely placed to the sound sources, their feeds will be less affected by environmental sounds and room effects. Alternatively, voiceover dialogue, sound effects, or Foley can be recorded separately and added to 3D space as point sources during post-production, which we will discuss further in the next section.

5.2.4 *Tracking*

In studio filming, the audio crew will often have access to a built-in sound system, where the audio captured on set will be routed to a control room either wirelessly or with cables through patch panels on set. In the control room, the microphone feeds will be tracked on a dedicated computer running a digital audio workstation (DAW). This is similar to how most audio recording studios are set up. In fact, soundproofed film sets are often referred to as "sound stages."

In location filming, however, the film crew will work with mobile recording equipment. The microphone feeds will be patched into a field recorder that encapsulates the features of a recording studio, including microphone preamplification, analog-to-digital conversion, monitoring, mixing, and storage. For a first-order Ambisonic microphone, a field recorder with at least four input channels is necessary. The output of the Ambisonic mic can be recorded into either four separate monophonic audio files or one audio file with four interleaved channels.

As we mentioned in Section 4.3.3, with Ambisonic microphones, it is essential to match the preamp gain for each of the four channels to ensure that the soundfield is captured evenly across all its dimensions. This rule of thumb should be exercised with both the preamps built into the field recorders used in location shooting and the outboard microphone preamps used in film studios. With preamps that offer analog gain control, it is possible for the levels of the individual channels to be slightly different even when the gain knobs are set to the same level. This is due to the inherent variations in tolerance in analog components, such as the potentiometers used in preamp gain knobs. Digitally controlled preamps, on the other hand, allow microphone gains to be set to an accuracy of a tenth of a decibel. For this reason, it is advisable to use a preamp with digital gain control to ensure a uniform level across all channels of the Ambisonic microphone.

If the production involves spot mics, there will need to be additional channels on the field recorder. Although the gain on these channels can be set independently, a sufficient amount of headroom should be left on each channel to avoid clipping and distortion during tracking. In the next section, we will discuss how an Ambisonic recording, captured either on set or on location, is processed and mixed with spot recordings in post-production.

Capturing VR Film and Its Implications for Audio

3D 360° cameras often rely on an array of image sensors with fish-eye lenses distributed on the circumference of a circle or the surface of a sphere. Although there are panoramic cameras with similar configurations that capture 360° photos and videos, a VR camera employs pairs of adjacent sensors to encode stereoscopic images that can address the left and right eyes individually, affording the viewer a sense of visual depth. Examples of modern spherical VR cameras include the Insta360 Pro and Kandao Obsidian Pro. There are also custom systems based on arrays of action cameras, such as the GoPro Hero and Insta360 One. Either way, a stitching algorithm is necessary to amalgamate the individual images from each sensor into a cohesive panoramic view. Therefore, most modern VR cameras come with dedicated stitching software for rendering 3D 360° photos and videos.

The spacing between the individual lenses of a VR camera is modeled on the human interpupillary distance (i.e., the distance between the center of each pupil). This is analogous to how binaural (dummy head) microphones, which we discussed in Section 4.2.1, are modeled on the distance between our ears. Since the VR camera and the Ambisonic microphone essentially function as the eyes and ears of the viewer, it makes sense to position them as closely as possible to avoid a spatial displacement between the audio and visual feeds. With that in mind, the blind spot of the camera, where the mounting equipment is located, can be a good place to position an Ambisonic mic. Nevertheless, occlusions that can be caused by the microphone's body, camera, or mounting gear should be accounted for while placing the microphone. Although the masking caused by equipment might be negligible, the path between the microphone and the primary sound sources in a scene should be as free from occlusions as possible.

In modern VR cameras, the tight placement of lenses across the surface of the camera can limit the size of the lens element and the image sensor associated with it. This limitation can impact the low-light performance and, therefore, the image quality of a VR camera. The type and extent of lighting are crucial for establishing a visual style in any kind of film production. However, unlike traditional film that can benefit from elaborate lighting setups, VR film can accommodate limited use of artificial lighting due to the panoramic view of the camera. In traditional film, the key and fill lights used for articulating the characters in a scene are often placed to the side of the camera right outside the frame lines. In VR film, such placement would put the lighting equipment directly in the audience's field of view. If a scene is set in an indoor area, the lights that are already built into this space, such as recessed and track lights, could be used as light sources that would not seem out of place in VR view. However, the luminous power of these utility lights will be significantly lower than that of studio lights.

When the story and production strategy permits it, the use of direct or indirect sunlight can be a valuable asset for VR film. The sun, which is the most powerful light source available to a filmmaker, can be integrated into a location shot in VR film without breaking immersion, since it is a likely object to encounter in an outdoors environment. A possible downside of using natural light is shifts in weather that could cause changes in the hue and intensity of the lighting in a scene. This can cause under or over exposure as well as visual continuity issues if a scene is shot over an extended period of time. Natural light can be leveraged in both open-air locations and indoor spaces with windows. While an outdoor shoot on a clear day may benefit from a substantial amount of natural light, it creates challenges for the audio crew in terms of dealing with unwanted ambient and wind noises. In traditional film, the boom operator would use a windshield to mitigate such noises. This often involves a zeppelin, which offers a foam layer and an enclosed air space around the shotgun microphone, and a dead cat, which is a fake fur layer dressed over the zeppelin for additional wind protection. Similar accessories for blocking wind are also available for Ambisonic microphones. In any case, each production will require unique technical and creative strategies to deal with such environmental factors. In Sections 5.5 and 5.6, we will explore case studies that utilize both studio lighting and natural light sources in a VR production context.

5.3 Post-Production Sound

A sizable portion of film sound is created after filming, during a phase known as post-production. In this phase, the audio material captured on set is supplemented and, in some cases, replaced with sounds that are captured in a studio or on location as well as those that are retrieved from existing audio libraries or synthesized from scratch. This phase is also when music is brought into the mix. Post-production is one of the areas where the existing workflows for film sound are fundamentally altered when working with immersive media.

5.3.1 Crew

The post-production phase involves a number of audio roles separate from those we find in a production sound crew. Two of the key creatives in post-production sound are the **sound designer** and the **Foley artist**. These play complementary roles in adding new sounds to the film besides those captured on set. The sound designer records, synthesizes, and processes sounds to be added to the film as sound effects (e.g., an explosion sound). The Foley artist, on the other hand, will perform sounds (e.g., the crunching of leaves as a character walks through a park) over the film.

In addition to these, post-production sound involves various audio editing tasks. The **dialogue editor** ensures that the dialogue recordings are audible, lined up with the film properly, and do not suffer from extraneous noises like pops and clicks. The **sound effects editor**, on the other hand, collects sounds from the production sound crew, sound designer, and Foley artist, and syncs these up with non-dialogue audio events in the scene. They may also gather additional sounds from existing audio libraries as needed.

The **composer** plays the critical role of creating the original soundtrack for the motion picture. This person is often selected by the director during the early planning phase of a film, and they work with orchestrators, musicians, and recording engineers to produce the soundtrack. The **music editor** plays a more flexible role between that of a sound editor and composer, aligning the recorded score with the film and applying necessary edits. The music editor also creates scratch tracks for film editors to work with while the score is being finalized. Besides original music written by the composer, film sound can also include preexisting music that complements the mood of the film. The **music supervisor** is in charge of curating and licensing this type of music. Once all of this audio material is gathered, the **re-recording mixer** makes sure they add up to a coherent whole. For instance, they may balance the dialogue with the music in a scene or adjust the level of synthesized sound effects versus those that are captured by the Foley artist.

While most of the roles in post-production sound are maintained in VR film, new creative and technical skills are required for working traditional film sound elements, like dialogue and effects, into immersive media. Modern DAWs are continually updated to offer new spatial audio features. Furthermore, there are new plugins and stand-alone software, such as the Dolby Atmos Renderer, that expand the audio capabilities of traditional production tools like DAWs and video editing software. Post-production tools like Adobe Premiere, Apple Final Cut Pro, and DaVinci Resolve are also updated with functionality to integrate spatial audio into film. While these changes introduce new elements to existing post-production workflows, they are likely to require all new roles dedicated to processing of immersive media in the near future.

5.3.2 *Preparing Immersive Audio for Post-Production*

With the production and post-production phases included, any given scene in a VR film might comprise a number of spatial media: a 3D 360° video captured on set or location, 2D or 3D graphics that may be overlayed on top of this video, an Ambisonic soundfield captured with a microphone on set or location, and an Ambisonic soundfield rendered with virtual point sources. These point sources can be spot recordings made during filming or sounds captured or generated during post-production. Similar to how a VR video can have a different visual resolution than the graphics added later, the soundfield captured on set can have a different Ambisonic order than the one rendered during post-production. The soundfield mics available today most commonly offer first- or second-order Ambisonics. However, the Ambisonic order of a soundfield synthesized from point sources (e.g., spot mic feeds, voiceover recordings, sound effects) is determined by the designer and can have higher resolution. Regardless of their resolution, all of these audio and visual media pertain to the same space and must be spatially aligned to work together. For instance, the video of a character speaking, the soundfield captured on set with an Ambisonic mic during this scene, and the close-up recording of the speech with a lavalier mic on the actor may all need to be rotated in space to originate from the same location in VR.

In Section 4.3.3, we discussed Ambisonic signal formats, starting with the four-channel A-format signal that represents the raw output from the four capsules of a first-order Ambisonic mic. We talked about how this signal was akin to a recording made with four separate microphones, where the individual channels are not spatially correlated. To derive an Ambisonic soundfield from these four recordings, the A-format signal needs to be converted into a B-format signal, wherein the channels form a cohesive whole spatially.

Most modern Ambisonic microphones bear markings that indicate the baseline orientation for a matching A-B converter software plugin. Based on these markings, we can configure the plugin to ensure that the orientation of the resulting soundfield matches that of the physical space where the recording was made. Figure 5.1 shows Sennheiser's Ambeo A-B Converter Plugin, which is designed to work with the company's Ambeo first-order Ambisonic microphone. On the left-hand side, we can monitor the four-channel A-format signal coming into the plugin. The controls in the middle let us specify how the microphone was positioned in recording space. The rotation dial controls which portion of the soundfield will be placed in front of the listener. When the dial is at 0°, the area in the recording environment that faces the "FRONT" marker on the microphone would be perceived directly in front of the listener. Regardless of the rotation of the microphone at the time of recording, the 3D soundfield can be rotated around the listener during post-production.

The position selection at the middle bottom of the A-B converter plugin allows us to specify how the microphone was oriented during recording. For instance, if the recording was made with the microphone hanging above the sound sources with the grill facing downwards, we would select the upside-down setting. In this scenario, leaving the position control in the upright setting would result in a vertically inverted soundfield.

Based on these settings, the plugin converts the A-format input into a B-format output. Much like the signal retrieved from the microphone, the B-format output consists of four channels. In the AmbiX format, these correspond to the W, Y, Z, and X dimensions,

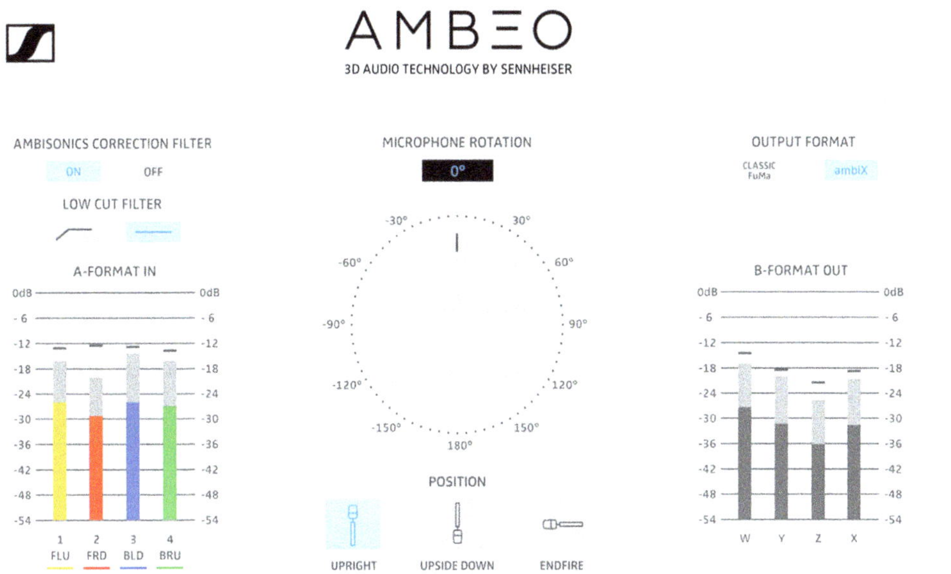

Figure 5.1 Screenshot of the Sennheiser AMBEO A-B converter plugin.

which are now spatially correlated and amount to a three-dimensional representation of the recorded environment. In a sense, the converter stitches the raw output from the four capsules into a cohesive soundfield, much like how a video stitching software merges the individual camera feeds from a VR camera into a 3D 360° video.

Besides the scene transformation controls, the format converter plugin offers three additional settings. The Ambisonics Correction Filter is described by Sennheiser as a specially tuned filter that improves the spatial accuracy of the plugin when paired with an Ambeo microphone.[4] The low-cut filter, which is sometimes built into microphones as an analog filter, can help eliminate some unnecessary low-frequency content, such as floor vibrations, electrical noise, or air handling rumble. Finally, the output can be switched between the FuMa and AmbiX exchange formats, which we discussed in Section 4.3.3. While there are legacy systems that are compatible with FuMa, most media platforms today support the AmbiX format. It is therefore advisable to use AmbiX to ensure greater compatibility with popular delivery platforms like YouTube and Meta.

Other soundfield microphone manufacturers offer similar plugins that are paired with their hardware. While these are often fine-tuned for the specific microphone that they are designed to work with, the parameter layouts of these plugins will be similar, if not identical, to the one we discussed here. There are also independent A-B format converter plugins that can work with any Ambisonic microphone. However, these can yield less accurate results than the manufacturer plugins due to a lack of microphone-specific calibration.

Once the B-format signal is retrieved from the raw input, it can be brought into a video editing software that supports Ambisonic audio. In Sections 5.5 and 5.6, we will describe post-production workflows for VR film, exploring how to mix monophonic sound sources with an Ambisonic recording and adjust their placement in space. We will also discuss working with modern audio and video editing tools to prepare immersive content for delivery. But before we dive deeper into the practical implementation of sound for VR film, let's discuss some of the more conceptual and aesthetic considerations in immersive media production.

5.4 Narrative Qualities of Sound in VR Film

In this section, we will touch upon some of the main areas where VR offers new opportunities for storytelling and the role of immersive audio in taking advantage of these opportunities. While most of these ideas are rooted in film theory and narratology, their application to VR brings about a technical and aesthetic reframing that particularly benefits from creative uses of immersive sound.

5.4.1 Diegesis

In Section 5.1, we categorized elements of film sound into dialogue, sound effects, ambience, and music. We can also categorize these sounds based on how they are situated in relation to the implied reality of a story. Imagine a flashback scene where the narrator recalls a childhood memory of Sunday breakfast. As the adult narrator's voice describes the moment, we see their younger self sitting at the kitchen table, enjoying breakfast with their family. The parents' dialogue in this scene would be an example of diegetic sound, as it belongs to the world of the story on screen. Similarly, the clinking of cutlery as the family eats their breakfast would be a diegetic sound effect. The voice of the narrator describing the scene, on the other hand, would be a non-diegetic sound. In this case, the narrator's voice is audible to the viewers but absent from the story world. A nostalgic musical score accompanying the narration would be another non-diegetic sound because this, too, would be inaudible to the characters in the scene. However, if the same music were played from a radio in the kitchen, it would be a diegetic element that exists within the story. The diegetic disposition of a sound shapes the world a viewer constructs from the story presented to them (Çamcı 2022). It can therefore impact the viewer's narrative experience and what they glean from the story.

The dynamic perspective that the viewer gains in VR has significant implications for the diegetic disposition of sounds in VR film. In traditional film, the viewer's perspective is fixed to that of the camera. Therefore, the filmmaker dictates the placement and orientation of the audience's point of view in the world that unfolds on screen. If a film has surround sound elements, these are arranged in reference to this point of view (e.g., the sound of an event occurring behind the camera would be heard from the rear speakers at the movie theatre). In contrast, VR film partially delegates the control of the camera to the viewer. While the camera's location remains locked to where the filmmaker placed it during production, its direction is mapped to the viewer's head-mounted display. This gives the viewer control over the framing of the scene at any moment. By moving their gaze through the scene, the viewer dictates not only what they see but also how they hear the virtual world around them.

When a sound maintains its position relative to virtual space as the viewer moves their head, it signals that the source of this sound is diegetic. Let's revisit the Sunday breakfast scene from earlier: the toaster in the kitchen goes off, making a sound that is spatially mapped to its location on film. If the viewer happens to be facing the toaster in that moment, they will hear the sound originating from in front of them. If, however, they are facing the other way, the sound of the toaster will come from behind them, likely steering the viewer's attention in that direction. The spatial adherence of this sound to its visual counterpart reinforces its diegetic disposition, indicating that the sound is grounded in the film world.

On the other hand, if a voiceover intended as non-diegetic narration was spatialized in the same way, it could suggest to the viewer that the source of this voice is also a part of the film world. As the viewer moves their head, the voiceover would maintain a fixed position in 3D space; despite the lack of a visual reference, the viewer could still pinpoint it among the other elements in the scene. This could lead the viewer to interpret the narrator as a ghostly presence occupying the same diegetic space as the characters and objects in the story. Unless this is the intention of the filmmaker, this could lead to narrative ambiguity. Previous literature has highlighted how distracting inaccurate spatial audio can be in VR films, emphasizing the importance of involving sound practitioners early in the planning process to prevent such issues (Butterworth 2022).

If the narrator is indeed meant to be non-diegetic, the voiceover could be presented as a non-spatialized (or head-locked) sound. As we discussed in Section 4.2.1, head-locked sounds are those that are anchored to the user's head, maintaining their position relative to the user rather than virtual space. Figure 5.2 shows a scenario where the VR user is presented with two spatial and two head-locked sound sources.

Whether sounds can be presented this way depends on the delivery platform's support for head-locked audio. For instance, YouTube supports VR video with up to six audio channels, where the first four are dedicated to first-order Ambisonics with two additional channels for optional head-locked stereo. Similarly, Meta VR's audio format combines second-order Ambisonics with head-locked stereo. These formats allow us to place the diegetic sources in the Ambisonic track, which responds to head tracking, while putting the narration in the stereo track, which would remain unaffected by head movements. As a result, the non-diegetic narration would maintain its position relative to the viewer rather than the film world. As the viewer moves their head, the voiceover would move

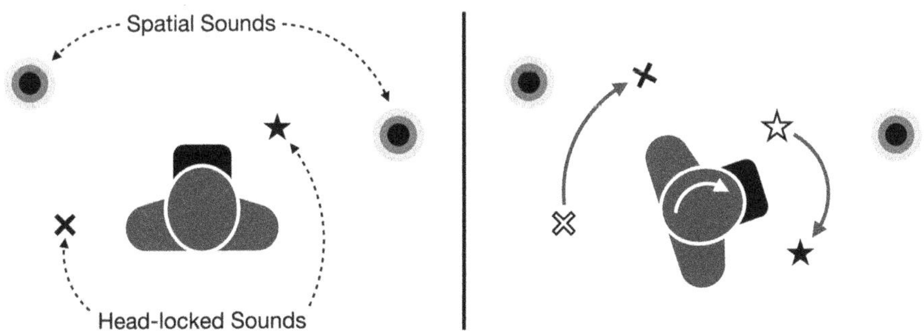

Figure 5.2 An overhead view of a VR user. On the left: the user is presented with two spatial and two head-locked sound sources. On the right: the user rotates their body to the right; while the two spatial sound sources maintain their position in virtual space, the head-locked sources move with the user, maintaining their position relative to the user.

with them, suggesting that it's an omnipresent sound that is external to the diegetic reality of the story. Following the narratologist Gérard Genette's terminology, this narration would be part of an extradiegetic layer outside the story world (Genette 1983). The extradiegetic layer itself could be a three-dimensional space surrounding the viewer. In other words, the voiceover could be placed binaurally anywhere around the viewer based on its narrative function (e.g., to the side as a whisper in the viewer's ear or to the back as a guiding voice). Either way, this space would be anchored to the viewer's perspective and remain detached from the story world.

If the delivery platform only supports an Ambisonic track without an additional head-locked component, all the sounds would be affected by the head tracking (i.e., maintain their positions in virtual space as the viewer moves their head). However, a source can still be made non-spatial by placing it at the origin. This would create a monophonic source perceived at the center regardless of head movement.

Film music requires similar considerations since it's often conceived as non-diegetic sound. If the delivery format supports a head-locked track, the music could be mixed for it. However, music often entails a greater degree of spatial detail than a less involved sound like voiceover. First of all, the music will likely incorporate multiple sound sources distributed across a supposed sound stage. On top of that, these may involve room effects captured during recording or added in post-production. Head-locking the music could cause a clash between the diegetic and extradiegetic spaces, as the head-locked space of the music would move with the viewer's head while the diegetic sounds remain fixed in the virtual world. This can be immersion-breaking depending on how the two are mixed and how prominent the music is in the overall sound.

To mitigate this, the music itself could be delivered in an Ambisonic format. As we will explore in our second case study, we can mix our music in Ambisonics from the beginning. If it is already mixed in stereo, the left and right channels could be positioned as Ambisonic point sources on either side of the listener. Either way, mixing the film music into an Ambisonic soundfield would allow it to be spatially anchored to the virtual environment, enabling the viewer to rotate their head within the music mix. Depending on the source, timbre, and dynamics of the elements in the music, as well as any room effects applied to them, the viewer might interpret the music as originating from the story world. Whether this affects the realism of a virtual scene will depend on the individual circumstances of the production. Even when we mix the music into the same Ambisonic soundfield as the diegetic sounds, we can apply reverb and room filters to set it apart from the rest of the film sound, thereby creating an auditory space that remains distinct from the implied universe of the film. Nevertheless, we must pay attention to the spatial relationship between diegetic and non-diegetic sounds and how it impacts a scene's plausibility in VR.

An interesting narrative strategy is to move elements of film sound between intra-diegetic and extradiegetic layers (Görne 2019; Çamcı 2022). Influential filmmaker Andrei Tarkovsky points to the use of electronic music to this end. Tarkovsky views instrumental music to be autonomous to the point of being a challenge to integrate as an organic part of a scene. However, he argues that electronic sounds can be more seamlessly absorbed into the sound world of a film. These sounds can then be manipulated to convey a character's interior world or reproduce their state of mind (Tarkovsky 1989).

Our previous research has shown that the diegetic disposition of a sound in VR impacts how the user interprets the VR environment and what decisions they make when they are presented with narrative choices (Çamcı 2019). For instance, a sound could be first presented as a non-diegetic sound without spatial cues or room effects, suggesting to the viewer that it is external to the implied universe of the story. This could be a voiceover track that narrates the scene. This sound could then be gradually imbued with localization cues that anchors it to a particular position in the scene. Such a change would transform the narrative function of the sound from representing an outsider who exclusively addresses the viewer to a story character that reveals itself as a part of the film world, subverting the relationship between the narrator and viewer.

5.4.2 Immersive Mise-en-Scène

"Mise-en-scène" is a French term that means "place on stage." It originates from theatre, where it's used for describing the arrangement of actors and props on the stage. In film, mise-en-scène is a general term that encapsulates everything that is placed in front of the camera. This includes the arrangement of actors, costumes, décor, lighting, and space. Film theorist Barrett Hodsdon describes mise-en-scène more generally as "the staging of action before the camera in a fictive context" (Hodsdon 1992), suggesting that movement of characters and objects can also be considered part of the mise-en-scène. Many members of the film crew from the director and cinematographer to lighting designer and production manager contribute to the mise-en-scène.

This concept takes on a new meaning in VR. As we have discussed before, the director's framing of a scene in the traditional sense is absent in VR film unless parts of the panoramic view are blocked out. As many film theorists would argue, the framing is an essential aspect of storytelling in film; the bracketing of the world outside of what's shown on screen is a means of narration, much like how a storyteller chooses which aspects of an event to communicate and which ones to leave out.

Camera placement and framing also allows the director to position the audience in relation to the story. These can either draw the audience into the narrative or position them as an outsider. Imagine a scene that shows a conversation between two characters. If the camera is placed at one of the characters' point of view, the other character would be addressing the audience as if they are a part of the story. On the other hand, if the camera is placed on the far side of the room, this could distance the audience from the story world or create the sense that they are eavesdropping on the conversation. While the camera positioning is critical to any kind of cinematography, the additional degree of agency that the viewer gains in VR places a different emphasis on it. This brings us back to the concept of diegesis and whether the camera placement situates the viewer as a diegetic or non-diegetic entity in any given scene. The placement of the camera will inform both the placement of the Ambisonic mic during production and the spatialization of additional sound sources in post-production.

Another narrative device in the director's arsenal is the movement of the camera. Together with framing, camera movement conveys a sense of action and guides the audience's attention. In VR, the control of this movement is partially handed over to the audience. With a panoramic view of the world around them, the audience can change the framing by moving their gaze across the scene. The VR camera could technically be moved on a dolly mechanism to give it a dynamic point of view. However, moving the user through virtual space while they are stationary in physical space can cause discomfort. Although a VR user's tolerance for virtual movement improves with experience, it's still a good practice to either keep the camera stationary or to only move it at a constant speed so as to avoid simulation sickness. This way, the experience can cater to a broad range of viewers with varying degrees of experience with VR.

That being said, as the consumer adoption of VR expands, the average experience level that a creator can expect from viewers will continue to increase. As a result, the best practices for camera movement in VR film might evolve over time. Nevertheless, a fixed camera position from where the viewer can look around can be leveraged as a creative constraint that informs our approach to immersive staging in VR film. This forces us to not only think critically about where we situate the viewer in relation to the story but also what the viewer will see or hear when they inevitably look away from the main subjects in the scene as they explore the immersive setting.[5]

The ability to observe the virtual space of a film in 360° encourages viewers to actively explore the scene. In doing so, they gain agency in the narrative and draw meaning from the mise-en-scène by shifting their attention across its different elements. This presents the filmmaker with the challenging yet intriguing task of striking a balance between creating a setting that justifies the immersive delivery and maintaining the viewer's focus on key aspects of the story. In the VR film *This Is Not a Ceremony*, the filmmaker Ahnahktsipiitaa presents true stories from systemic challenges that First Nations communities have faced in the recent history of Canada.[6] Instead of actively engaging in the story as a character, the viewer is situated to bear witness to the struggles of the Indigenous people (Bédard 2023). One of the segments of the film tells the story of Brian Sinclair, an Indigenous Canadian who passed away from a treatable condition after being left unattended for thirty-four hours in a hospital waiting room. In a powerful example of immersive mise-en-scène, the viewer is placed in the middle of the waiting room, while a non-diegetic narrator recounts the timeline of the events leading up to Sinclair's death. The scene is populated with healthcare workers and other patients with Sinclair being the closest to the viewer. The tension between the non-diegetic narration that foreshadows Sinclair's fate and the inaction that the viewer witnesses in the diegetic world of the film instills a feeling of desperation. When Sinclair eventually passes away, exploring the immersive setting and observing the reactions of various characters in the scene creates a haunting experience.

5.4.3 Non-linear Storytelling

While this chapter primarily focuses on fixed-media film production, a number of VR films leverage interactivity to tell non-linear stories, where the viewer can take further agency in their experience. This can include interacting with objects in a scene or even influencing the progression of events in the story. Non-linear storytelling is explored more commonly in animated films since computer graphics make it easier to implement interactive features. However, there are also VR films that follow a choose-your-own-adventure model where the viewer can pick between different possible scenarios. For instance, in the VR film *Afterlife*,[7] the filmmakers present the viewer with 29 choices that lead the story to different endings.

Non-linear films can demand greater awareness from the viewer by way of involving them in the story's progression. The viewer may need to pay attention to the elements of the panoramic scene or details of the story that could inform their decision when prompted to make a choice. In these scenarios, audio not only serves the typical diegetic and non-diegetic functions that we encounter in traditional cinema but also acts as an additional channel of information that can help the viewer navigate the interactive experience. It can guide the viewer's attention to specific points in the panorama, provide auditory feedback when they interact with the scene or a non-diegetic user interface, and signal transitions between forking paths in the story.

Animated VR films that leverage computer graphics and interactivity blur the distinction between films and video games. Indeed, video games have a long history of offering story-driven cinematic experiences. Furthermore, gamification finds broader use in VR in a range of fields from education to physical therapy. In Chapter 6, we will take a more detailed look at the role and implementation of sound in non-linear and interactive VR experiences such as games.

5.4.4 Heightened Reality

> ...when Bergman uses sound apparently naturalistically—hollow footsteps in an empty corridor, the chime of a clock, the rustle of a dress, the effect is in fact to enlarge the sounds, single them out, hyperbolise them...He singles out one sound and excludes all the incidental circumstances of the sound world that would exist in real life.
>
> (Tarkovsky 1989)

Film sound often seeks to capture "an idealized reality, a world carefully filtered to eliminate sounds that fall outside of understanding and significance" (Belton 1985). Elements of film sound might be balanced differently from the sounds we encounter in the real world to enhance clarity, create dramatic effect, or convey story details. For instance, when a film character is being stealthy, the sounds emanating from their interactions with objects or environments might be mixed down to a level far below their real-life volume. Conversely, sounds can be mixed louder than they actually are in order to draw attention to them. This could be the sound of a leaky faucet that's meant to irritate a character or the sound of footsteps signaling the arrival of an adversary.

In VR film, the first-person perspective of the viewer can make such divergences from reality more noticeable. The ability to move one's point of view allows for greater scrutiny of the sound design in a scene. If a scene, for instance, establishes a realistic auditory space with reasonably leveled character dialogue and a convincing room response, any exaggerated or understated sound can interfere with the environment's plausibility. The viewer might confuse diegetic elements with non-diegetic ones and have a hard time placing a sound source in the virtual space that surrounds them. On the other hand, this type of heightening or subversion of an auditory reality can serve a narrative function in VR. For instance, mixing a sound at a lower level lower than the audience would expect can compel them to turn an ear towards its source to hear it more clearly. This can promote an active engagement with the scene or even help reveal visual features that are outside the viewer's visual field.

Apart from using amplified sound to heighten reality, traditional cinema also employs audio effects to evoke specific sensations in the listener. For instance, reverberation is often used to suggest a dream state, while low-pass filtering can indicate that a character is feeling unwell or dissociated from their surroundings. While these effects are also applicable to VR film, the designer should consider how these might interfere with the viewer's perception of virtual space, as both reverberation and filtering contribute to a room's acoustic characteristics. In VR, where the viewer's perspective is first-person and dynamic, such effects may need to be exaggerated to prevent them from being mistaken for spatial or localization cues. In fact, exaggerated sound-image interactions have been described to facilitate the "ebb and flow of tension and aesthetic release" in cinematic VR (McArthur, Stewart, and Sandler 2017).

Another method to heighten reality through sound in VR film is to deviate from spatial fidelity through the use of the ventriloquist effect, which we touched upon in Section 4.1. In this effect, our perception of a visual source "captures" the sound source associated with it, causing us to perceive the sound as if it's emanating from the visual source even though the two stimuli do not overlap in space (Alais and Burr 2004). The effect is named as such because the same principle underlies how we perceive the voice of a ventriloquist as originating from their puppet. In VR film, this effect can be utilized to emphasize critical dialogue by gradually moving it from a fully spatialized sound to head-locked mono sound. This ensures that the dialogue is clearly delivered to the user while the ventriloquist effect helps them associate the mono sound with its visual source in the scene. It is worth noting that this works more effectively in 3D 180° film, where most of a visual scene is within the user's field of view. In 3D 360° film, where the user can turn their back on the visual source of a dialogue, the perceived association between a head-locked sound and its visual source via the ventriloquist effect can be disrupted.

5.5 Case Study 1: Producing a VR Film

In this and the following section, we will describe two VR film projects that we carried out and documented from start to finish for this book. While the two projects share some tools and techniques, we strived to follow two different workflows to demonstrate distinct creative approaches. In our first case study, we will discuss the making of a VR film from pre-production to post-production with a focus on the recording and processing of immersive sound. We will describe the aesthetic and technical considerations that went into the project and detail its implementation. The post-production workflow will involve mixing VR audio in a film editing software, namely Adobe Premiere.

For this project, we collaborated with the filmmaker and composer Andy Kirshner to produce a segment of his upcoming documentary art film *Sex Radical*. The film explores the cultural shifts and rise of feminism in the Victorian Era through the life of Ida Craddock, who was a nineteenth-century free speech and women's right advocate. Here is Kirshner's description of the project from the film's website:[8]

> At the turn of the last century, an act of Congress made it illegal to send any information about sex, contraception, or abortion through the U.S. mail. But an iconoclastic feminist, spiritualist, and sex educator named Ida Craddock publicly challenged that law—and the law's chief enforcer, Anthony Comstock. Defending the right of a woman to "control her own person," she risked everything. In an era when religious conservatives make no secret of their plan to use the 150-year-old Comstock Act to limit reproductive rights, Ida Craddock's story is inspiring, illuminating, and timely.

5.5.1 Production Preparation

There were several phases to our preparation for the film. The early planning involved meetings with Kirshner to discuss story ideas and whether these would be suitable for a

VR film from both narrative and technical standpoints. With each idea, we evaluated how the immersive presentation contributed to the story, how the audio and visual elements would be situated around the audience, and which locations could be utilized for the shoot. We also discussed equipment needs (i.e., camera, microphones, lighting) and the limitations that immersive recording could impose on the production. After evaluating several possibilities, Kirshner proposed shooting one of the scenes from *Sex Radical*, for which he was doing early planning at the time.

The scene is shot in a single seven-minute take in the style of a theatrical play. In the scene, four characters seated at a round table engage in a seance, where they attempt to contact the dead. We found this idea fitting for a VR film for several reasons: the placement of the characters in the scene and the dialogue progression between them gave us several options for mise-en-scène. Besides the characters, the Victorian setting of the story would allow us to add props and set dressing that could make the panoramic view more engaging. Moreover, the paranormal nature of the story could create opportunities to experiment with uncanny or abstract sonic and visual elements. Finally, the indoors setting of the story would allow us to achieve a more controlled production environment, where we could leverage the staging to hide crew behind walls during the shoot.

We scouted several locations on the University of Michigan campus, including libraries and ballrooms, that could serve as a Victorian backdrop for the film. Ultimately, we decided to work at the University of Michigan's Duderstadt Center Video Studio, which is a black-box production space wherein the seance room could be built from the ground up. This gave us more control over the setting and scheduling. This also simplified the technical implementation of the project since the production space had built-in lighting and audio recording capabilities. A few months ahead of the filming, we met with the production crew of the Video Studio to discuss creative goals, technical needs, and production timeline.

5.5.2 Mise-en-Scène

Remembering our conversation on mise-en-scène in Section 5.4.2, it is easy to understate the importance of staging of characters and objects in a VR scene. Giving the audience the freedom to choose what they pay attention to in any given moment brings up an interesting creative challenge that urges us to reimagine some foundational aspects of filmmaking and, more broadly, storytelling.

To account for the fully panoramic view of the VR camera, we decided to set up a whole room with walls and curtains with a round table in the middle as shown in Figure 5.3. The production designer, Laura King-Pazuchowski, decorated the background with furniture and picture frames. In the spirit of a seance, the room was dimly lit with a softbox array in the ceiling. To further highlight the actors and give them an eerie look, we placed small LED panels on the table to light their faces from below. As shown in Figure 5.4, the lights were placed behind candles on the table so that they are not directly visible to the viewer.

Given the story at hand, we had a few options for where to place the camera and, by extension, the viewer. Situating the viewer as a participant in the seance would entail setting up an extra chair for the camera and having it placed around the head height of the actors. However, since the film did not involve any interactivity, placing the viewer as a character in the scene without any agency besides the ability to look around did not seem like an ideal option. We also considered moving the viewer a little further back to

Figure 5.3 Director Andy Kirshner (standing center, behind the VR equipment) discusses the scene with the actors around the seance table.

Figure 5.4 Close up of a small LED panel used for highlighting an actor. We placed the panels behind candles on the table to hide them from the camera's view.

give them an outside view of the seance, but since the story became primarily driven by the dialogue between the characters, distancing the viewer from the conversation was not preferable either. Ultimately, we decided to place the viewer at the center of the table as seen in Figure 5.5. While this placement gives the viewer a ghostly presence in the scene,

Figure 5.5 The VR camera and the Ambisonic mic above it are suspended from the ceiling with strings over the middle of the seance table. The camera is aligned with the actors' heads when they are seated around the table. Note that these pictures were taken under work lighting.

it provides a unique perspective into the story with equal emphasis on each character. This encourages the viewer to pay attention to the characters not just when they are speaking but also as they react to the other characters. This also places the viewer at the very center of the room, prompting them to explore the immersive staging of the scene.

5.5.3 Production Setup

On the day of production, we worked with the camera, lighting, and audio crews to set up the VR equipment for filming. Having decided to position the camera at head level towards the middle of the seance table, we had two options for mounting. We could either place the camera upright on a tripod on the table or suspend it upside down from the ceiling. We followed the latter option, as shown in Figure 5.6, for several reasons. First, the table in the scene was a round wooden table with a 5 ft. (152 cm) diameter. Despite being stable, the table could still transmit vibrations when the actors put their hands on it. While the gyroscope built into the camera could account for changes in the rotation of the camera, vibrations would also result in vertical or horizontal displacements. Even subtle movements of the camera could cause discomfort in VR since they wouldn't match the viewer's head movements. Suspending the camera and the Ambisonic mic from the ceiling allowed us to isolate the equipment from any stage vibrations. Second, since there were ambient light sources placed on the ceiling, we would already need to block out this area of the scene in post-production. By suspending the camera upside down, we aligned the ceiling with the side of the camera where the mounting equipment was located.

Additionally, suspending the Ambisonic mic prevented it from being placed in the narrow gap between the table and the camera, which could have caused unwanted early reflections in the recording. Finally, while the production designer was preparing the stage, we decided that the fabric and props they set up on the table would be interesting to showcase in VR. Therefore, we wanted to avoid obstructing these elements by setting up the mounting equipment on the table.

The "FRONT" marking on the microphone was lined up with one of the lenses on the camera to make it easier to correctly align the soundfield with the panoramic video in

Figure 5.6 A close up of the VR camera and the Ambisonic mic suspended from the ceiling with wires.

post-production. Once the camera was secured in place, we ran a stitching calibration through the Insta360Pro mobile app. While this calibration does not affect the recorded video, it optimizes the real-time stitching, which is useful for monitoring the video during the shoot. The monitoring is also achieved through the Insta360Pro mobile app.

We equipped the four actors with lapel mics to capture their dialogue individually. Modern lapel mics are small enough to be attached to clothing inconspicuously. The wireless transmitters for these microphones can also be hidden under clothing. The individual feeds from the four wireless lapel mics and the four channels of audio from the Ambisonic mic were routed to dedicated preamps. As we mentioned in Section 5.2.4, for the B-format Ambisonic signal to be properly retrieved from the A-format output of the Ambisonic mic, all four channels should be at the same gain. To achieve this, we used a microphone preamp with digital gain controls, which allowed us to match the gain on the Ambisonic channels with high precision.

In the scene, the characters not only engage in dialogue but also sing, during which their voices become much louder than when speaking. With this in mind, the gain levels on the preamp were set low enough to accommodate these fluctuations in loudness. The preamp outputs were routed to a DAW for tracking. Each take was marked with a clapboard on the "FRONT" side of the Ambisonic mic so that we could synchronize the audio and video signals during post-production. Most VR cameras also have built-in regular or spatial mics that record directly onto the video. Although the sound captured through a camera's built-in microphone will be of lower quality than recordings made with dedicated microphones and preamps, it can still be useful in post-production as a reference for synchronizing external audio tracks to the video.

5.5.4 *Post-production*

The first step in VR post-production is processing the media captured by the Ambisonic mic and the VR camera. As we described earlier in this chapter, these devices record the scene around them using a spatial configuration of sensors (i.e., microphone capsules and camera lenses). The individual feeds from these sensors are not immersive on their own, so we need to retrieve the 3D scene by combining these feeds. For audio, this involves processing the A-format signal from the Ambisonic mic (i.e., the independent recordings from the four capsules) into a B-format signal (i.e., a first-order Ambisonic soundfield). For video, a similar process entails stitching together the content captured by the adjacent lenses on the VR camera into a fully panoramic 3D view. We then combine the processed media in video editing software and introduce additional audio and visual elements.

5.5.4.1 *Processing Spatial Media*

To retrieve the B-format Ambisonic signal, we use REAPER,[9] a DAW with a generous trial period[10] and a relatively low-cost license. Since REAPER was one of the first DAWs to support multichannel tracks for immersive audio processing, there are numerous online resources on how to work with Ambisonic signals in it. While REAPER's accessibility and robust multichannel audio features make it a good choice for VR audio applications, many modern DAWs also support the processing of Ambisonic signals. Therefore, the techniques demonstrated here should be applicable across different platforms, though there may be slight differences in user interface and workflow.

Figure 5.7 illustrates the signal flow starting from the individual capsules of the Ambeo mic. These are labeled as 1) FLU (Front Left Up), 2) FRD (Front Right Down), 3) BLD (Back Left Down), and 4) BRU (Back Right Up). The outputs of these capsules are captured on four mono tracks in the DAW. Since the A-B Format Converter plugin has four inputs and four outputs, the mono tracks are then routed to a four-channel track that has this plugin. The output of this track will provide the B-format signal, which we will import into the video editing software.

Figure 5.8 shows a segment of a REAPER session that implements this process. There are four mono tracks dedicated to the individual feeds from the Ambisonic mic's four capsules, routed into a four-channel Master track. In the FX chain of this track, we add the Ambeo A-B Format Converter plugin, which we previously showed in Figure 5.1. Note that the order of the incoming mono channels must match the input channel order specified in the plugin. Since the Ambeo was hanging from the ceiling upside down, the position setting needs to be adjusted accordingly. On the right-hand side of Figure 5.8, we see the routing settings for the third track with its output routed to the third channel of the Master track. Although the routing windows for the remaining tracks are not pictured here, they are similarly configured based on their order. The figure also displays the output settings for the Master track, with its channel count set to four to accommodate the four-channel A-format input and the B-format output of the converter plugin.

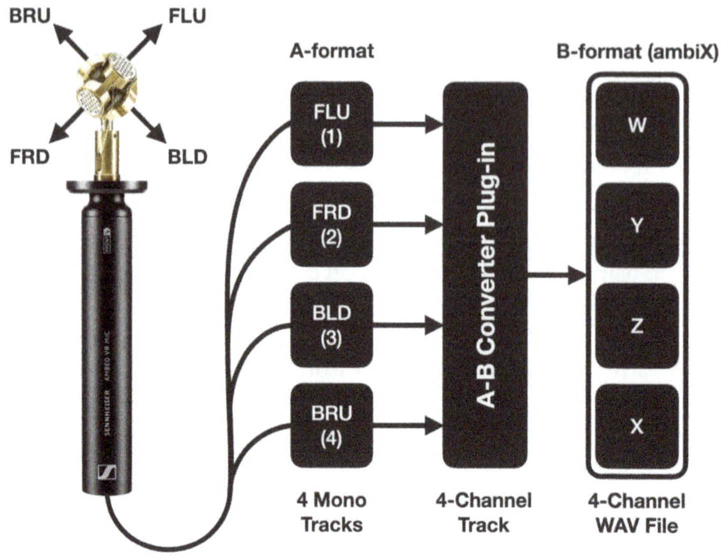

Figure 5.7 Channel routing from the Ambisonic mic into the individual mono tracks corresponding to the FLU (Front Left Up), FRD (Front Right Down), BLD (Back Left Down), and BRU (Back Right Up) capsules of the mic. Note that the capsules are labeled from the mic's perspective. The mono tracks are then routed to a four-channel track with the A- to B-Format Converter plugin. The output of this track is then rendered into a four-channel audio file.

Next, we apply a similar process to the VR video. Each of the six lenses on the Insta360 Pro camera captures a monoscopic video that is not unlike a video recorded with a traditional camera. These six videos then need to be stitched together to create a panoramic view, which can be either 2D or 3D. In 2D mode, the panorama is divided into six segments, with each lens covering a specific portion of the view. In 3D mode, adjacent cameras act as pairs of eyes, resulting in two panoramic views—one for each eye—that are composited into a single video file. The compositing process arranges the two panoramas either side by side or over and under. The choice between these two options depends on the post-production software and delivery platform. Most video editing tools with VR features allow the user to specify how the 3D 360° video has been composited. Therefore, as long as the settings are consistent, the video should appear as intended. In our case, we composited the left and right-eye panoramas using the over-under method, with the left eye's feed on top. A frame from the resulting video is shown in Figure 5.9. However, based on your intended delivery platform, you may prefer to select the side-by-side option, where the left and right eye feeds are lined up horizontally. In circumstances where you are targeting multiple platforms with different settings, you can also work in one of the formats and convert the over-under video to side-by-side (or vice versa) later.

While modern stitching software is generally effective at combining the feeds from different lenses, minor artifacts may still occur where two adjacent videos are merged. These artifacts can be monitored during production to avoid placing important subjects at stitch lines. However, non-real-time stitching software used during post-production can achieve a much more seamless video than what a real-time monitoring software can

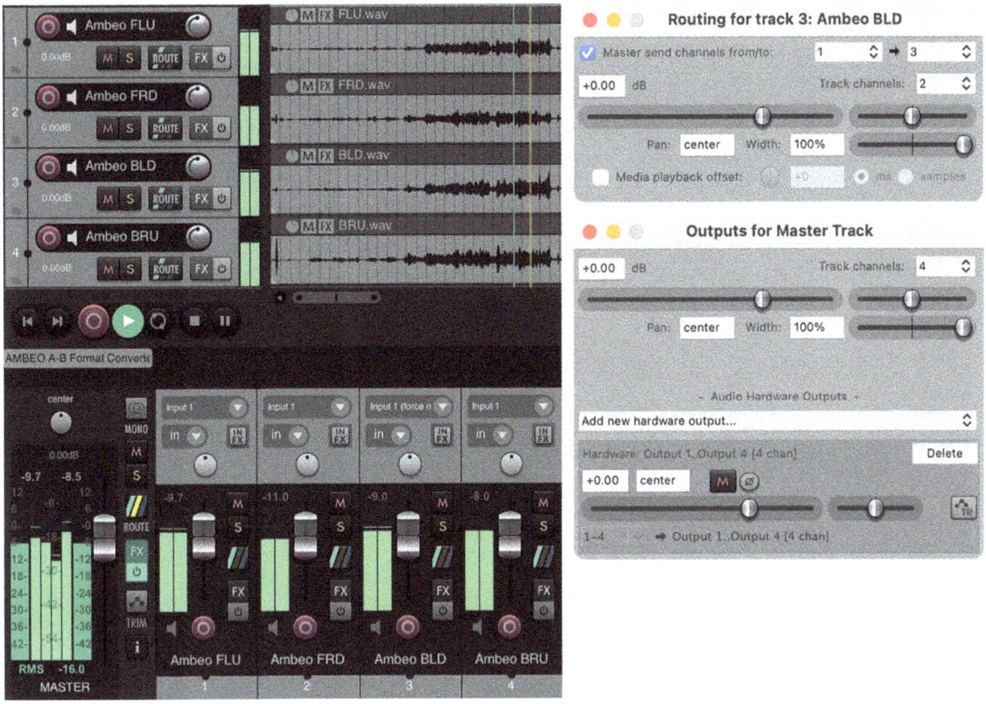

Figure 5.8 Screenshots from the REAPER project. The four mono tracks correspond to the four channels of the Ambisonic microphone. These are routed to the Master track, which has the Ambeo A-B Format Converter plugin. On the right hand side, we see the routing settings for Track 3 and the output settings for the Master track.

offer with today's technology. Stitching artifacts are often more pronounced at the camera's zenith and nadir, where all the videos converge. Fortunately, these points are at the very top and bottom of the VR video respectively, typically staying outside the user's field of view.

Figure 5.9 already reflects a couple of image corrections that we applied in video editing. In the previous section, we mentioned that the scene was lit from the ceiling using a softbox. To block this light from appearing in VR view, we created a solid-black image with gradient opacity to be superimposed on the VR video in Adobe Premiere. As seen in Figure 5.10, the gradient image is scaled so that it blocks the light but does not clip into the area where there are actors and decor. This way, the blocking image gradually fades away without interfering with the scene. Additionally, we applied color correction in Premiere to make this transition appear more seamless. We should note that the flattened panoramic image in Figure 5.9 overemphasizes the nadir and zenith portions of the video. Therefore, it may appear that a large portion of the scene is black. However, when this image is projected onto a sphere for VR view, the middle portion takes up most of the user's field of view, which is wider in the horizontal axis than it is in the vertical axis.

While there are other steps to video post-production, including editing, noise reduction, further color and image corrections, and the addition of graphics and titles, these are techniques that are more generally applicable to filmmaking and, therefore, lie beyond

Figure 5.9 A panoramic frame of the VR film in over-under format, displaying the four actors seated around the séance table holding hands. The top half shows the left-eye view, whereas the bottom half shows the right-eye view.

Figure 5.10 Application of the gradient image to block the softbox lighting seen at the top of left and right eye views.

the scope of this book. There is extensive prior literature which offers a more thorough documentation of video post-production (for a few examples, see Barnwell 2019, Case 2013, Silverman 2004).

5.5.4.2 Audio Mixing

Now that we have retrieved the spatial media from our recordings, we can bring them together in Adobe Premiere. Note that an audio post-production specialist will likely use a DAW, rather than a video editing tool like Adobe Premiere, for mixing film sound. However, since this particular project involved a relatively small number of audio tracks, we will use it as an opportunity to showcase the Ambisonic audio workflow in a video editing tool. In our second case study, we will utilize a DAW to mix the sound for another VR film.

The audio tracks in this project include the first-order Ambisonic recording, the mono recordings from the spot mics on the four actors, sound effects, and music. Figure 5.11 shows the Premiere project for the film. The current layout of the editor prioritizes audio mixing with the Audio Track Mixer panel on the top left. The channel strips shown here correspond to the individual audio tracks. The timeline for these tracks are displayed in

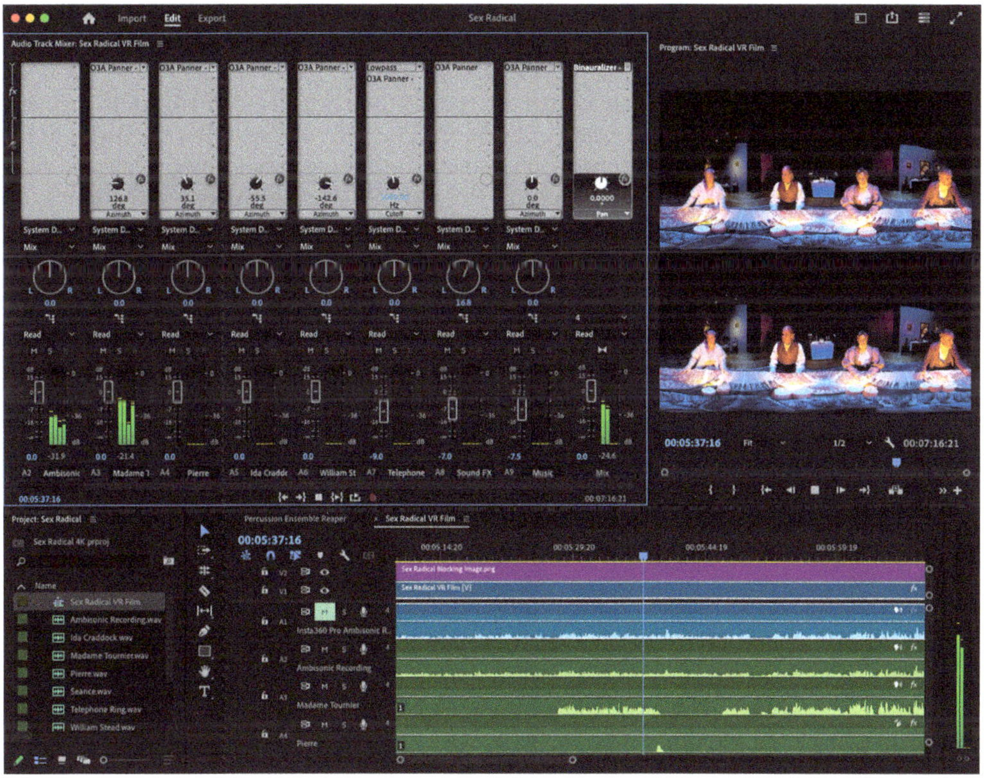

Figure 5.11 Adobe Premiere Session with the Audio Track Mixer on the top left, Program Monitor showing the over-under video on the top right, Project Folder on the bottom left, and the Timeline panel on the bottom right.

the bottom right panel. While the top two tracks correspond to the VR video and the light-blocking layer, the ones below these represent the audio tracks. The "Insta360 Ambisonic Recording" track in the Timeline panel has a highlighted button with the label "M," which indicates that this track is currently muted. This is the first-order Ambisonic recording from the built-in mic of the VR camera. As we mentioned earlier, this track functions as a useful reference for aligning audio recordings with the VR video. However, it is a lower-quality recording than the one we made with the Sennheiser Ambeo, so we keep it muted.

Looking at the audio tracks in the Timeline panel, you may notice that each track is labeled with "4." This represents the number of channels in each track. In Premiere, there are four types of audio tracks: Mono for monophonic files, Standard for stereophonic files, 5.1 for surround sound, and Adaptive for other configurations, including Ambisonic audio. For instance, when we drop a four-channel first-order Ambisonic file onto the timeline, Premiere automatically creates a four-channel Adaptive audio track. New tracks can also be added from the Sequence menu or by right-clicking on one of the existing tracks and choosing "Add Tracks..." from the pop-up menu. Once an audio file is added to the Timeline, Premiere will use the number of channels in this file as a reference when creating new Adaptive tracks. For instance, when we added the VR video to the timeline, it came with the four-channel Ambisonic recording from the built-in mic. This effectively sets the default number of channels for Adaptive tracks to four. However, the number of channels in the audio tracks can be changed from "Sequence > Sequence Settings... > Audio > Number of Channels." Given the limited number of sound sources, which are stationary and evenly distributed across the panorama, we decided to set the output audio format of the film to first-order Ambisonics. This not only matches the order of the recording made with the Ambeo mic but also ensures compatibility with delivery platforms like YouTube VR. Therefore, we set the channel number of the audio tracks in Premiere to four.

Mixing in B-format Ambisonics would typically require an Ambisonic speaker array, which may not be accessible to all of our readers. To mix an Ambisonic recording with headphones during post-production, we can convert it to binaural audio. To achieve this, we will use Premiere's built-in Binauralizer plugin on the Mix track. Once we are done with the mix, we will need to remove this plugin to export the film with first-order Ambisonic audio. When the film is being viewed in VR, the delivery platform will apply a similar process to convert the Ambisonic signal to binaural in real time. In this case, however, the binaural rendering in the VR headset will rely on head tracking. By contrast, if we export the film with binaural rendering—rather than as an Ambisonic signal—the audio will be head-locked and won't respond to head movements.

While the Ambisonic recording is already converted to B-format, we still need to encode the monophonic spot recordings from the lavalier mics into first-order Ambisonics as point sources. To achieve this, we will use Blue Ripple Sound's O3A Panner, which is part of a free suite of plugins that can process up to third-order Ambisonics.[11] There are multiple versions of the panner plugin with varying functionality. In this case, we will use the O3A Panner (Large) shown in Figure 5.12. This version of the plugin allows us to load a frame from the video and position the sound source using a cursor on the image. Moving the cursor adjusts the Azimuth and Elevation values for the sound source. The plugin shown in Figure 5.12 is attached to the audio track for the spot recording of the third actor from the left with the cursor aligned with the actor's mouth. We then apply

Figure 5.12 O3A Panner (Large) Plugin for Ambisonic panning. A frame from the film is loaded as reference for moving the sound source across the scene. Note the crosshairs over the actor second from the right. This indicates the current sound source location, which is reflected in the azimuth and elevation parameter controls as well.

the panner plugin to the remaining spot recordings and adjust their positions in the scene. If an actor moves in the scene, the cursor position would need to be moved with it. In Premiere, this can be achieved by keyframing the plugin parameters to track the actor's position.

Besides the dialogue, the scene includes the sound of a phone ringing, which was not recorded on set. When the phone rings, the psychic gets up and goes to an adjacent room off-screen to answer it. Even though the phone is not visible to the viewer, it is still a diegetic element in the story. We therefore spatialize it based on where the character leaves the scene. For the phone ring sound, we found a sample of an old phone on free sound.org. Since the phone is in a room adjacent to the one shown in the scene, its sound would be muffled due to the wall between the two rooms. To achieve this effect, we applied a low-pass filter to the sample in Premiere. In traditional mixing, effects like EQ, compression, and delay may be added after a sound is panned. When using regular audio effects plugins in an Ambisonic workflow, these effects would need to be applied before the sound is encoded into an Ambisonic B-format signal. This is primarily because most audio effects are designed as mono or stereo plugins that cannot handle the number of channels in an Ambisonic signal. However, there are multichannel plugins, such as IEM Multi EQ and IEM MultiBandCompressor,[12] that have enough inputs for up to seventh-order Ambisonics and can therefore be used after a mono signal is encoded into B-format.

The project includes another track for sound effects that mark critical moments in the story, such as when the psychic is taken over by a spirit, or when the ghostly figure of a woman materializes. To highlight these moments, we used abstract sounds performed on cymbals and synthesizers. Although these sounds are meant to be non-diegetic (i.e., inaudible to the characters in the scene), we leveraged them to guide the viewer's attention to key elements in the story. For instance, the cymbal sound that marks the materialization of the ghostly figure is mapped to where that imagery occurs. The abstract nature of the sound and the reverberation applied to it suggest to the viewer that the sound originates from an extradiegetic layer separate from the world on screen. However, its spatialization directs the viewer's gaze to a specific point in the visual field.

Finally, the project has a music track, which consists of percussions that function as the main musical theme of the scene and an ambient drone that establishes an eerie mood during parts of the seance. While a small amount of reverb has been applied to the music track, it is positioned at the very center of the scene. Therefore, the music is head-locked to the viewer and does not maintain a position relative to the scene. This configuration reinforces the nondiegetic role of the music. A similar effect could be achieved by addressing the head-locked tracks of the delivery platform. For instance, the last two channels of a 6-channel upload to YouTube are processed as a head-locked stereo track. As we mentioned in the previous section, another possibility would be to mix the stems of the music track into the soundfield as point sources to have the music be anchored in virtual space. However, since the music in this scene is relatively minimal, there weren't too many elements to distribute around the viewer. Nevertheless, we will explore the Ambisonic mixing of a track with multiple channels in our second case study.

Aligning Sounds and Visuals

A soundfield recording can be spatially misaligned with the VR video depending on the camera's orientation relative to the Ambisonic mic during production. In other words, the front of the camera may not correspond to the front of the captured soundfield. This would require rotating either the visual or the audio scene to match the other. Most A-B format converter plugins offer controls to rotate the soundfield captured with Ambisonic mics. Additionally, there are soundfield transformation plugins, such as O3A Rotation and IEM SceneRotator, that can rotate a soundfield as a whole. Alternatively, we can keep the soundfield in its original orientation and rotate the visual scene to match it. This can be achieved by applying a VR Projection effect to the video in Premiere and adjusting the "Pan" parameter to rotate the view around the vertical axis.

5.5.5 Exporting

The final step to post-production is to export the film in the intended media delivery format. While Premiere already offers exporting presets for VR video, the media format can also be set up manually based on the settings recommended by the delivery platform. In the current Premiere workflow, the video settings in the export window have a dedicated VR video tab, where the user can indicate whether a video is intended for VR, whether it's monoscopic or stereoscopic with over-under and side-by-side options, and what the field of view is. For a 3D 360° video such as ours, the field of view would be 360° horizontally and 180° vertically. Similarly, the audio settings will require us to set the number of channels to four and indicate that the audio is Ambisonic. Once the video file is exported, there may be additional steps based on the delivery platform. For example, YouTube VR currently requires the use of a stand-alone meta-data injector application for the video to be recognized as a VR video when uploaded to YouTube.

5.6 Case Study 2: Producing a VR Concert

For our second case study, we captured a music performance in VR to explore different aspects of VR film production than those we covered in our first case study. For instance, we wanted to incorporate musical acoustics and mixing, which led us to use a different set of recording techniques and post-production tools. We also wanted to film on location, rather than on a sound stage, to employ natural lighting instead of studio lights and a mobile recording setup. To this end, we collaborated with the Percussion Ensemble at the University of Michigan to record a performance of the composer Carolyn Chen's *Hamlet* for six percussionists. The piece is a reinterpretation of William Shakespeare's famous play. The performance features a variety of percussion instruments, including timpani, marimba, glockenspiel, and steel drums, as well as non-musical objects like a watermelon, toy robots, and kitchen knives, alongside spoken words. The project files for this case study can be accessed at vraudio.org under Chapter 5.

As we will further detail in the next section, Chen's score for the piece already situates the performers in an elliptical configuration where they face each other. This spatial layout, paired with the timbral and narrative richness of the piece, made it appealing for a VR concert recording. Furthermore, the transient nature of percussive sounds emphasizes their localization, making them particularly engaging to experience in an immersive context.

The recording was carried out in Hankinson Hall at the University of Michigan's School of Music, Theatre & Dance. This is a rehearsal hall with a 3,200-square-foot floor space with a thirty-foot ceiling. In one corner of the hall, there are windows that extend from the floor to the ceiling, letting in a substantial amount of natural light. The walls have fixed acoustic panels as well as curtains that can be used for attenuating some of the wall reflections, making the space less reverberant.

5.6.1 Production Preparation

In preparation for the recording, we met with the director of the Percussion Ensemble, Ian Antonio, to discuss the piece, location, and technical requirements for the VR shoot. We also met with the ensemble in Hankinson Hall to review the instrument layout and to scout the space for camera and microphone placement. During that meeting, we also briefed the performers on what to expect from the day of the recording as well as the eventual outcome of the project. We then devised a plan for mic selection and placement, signal routing, and gain staging. We also scheduled the setup, recording, and strike times. Since the piece is approximately 30 minutes long, we split the recording time into 15-minute takes to prevent the VR camera from overheating.

In this project, we wanted to compare two Ambisonic microphones, namely the Sennheiser Ambeo, which we also used in our first case study, and the Voyage Audio Spatial Mic. Whereas the Ambeo is a four-channel first-order Ambisonic microphone, the Spatial Mic is an eight-channel second-order Ambisonic microphone. Although a second-order Ambisonic signal consists of nine spherical harmonics (and, therefore, nine channels), the Spatial Mic, as well as other octahedral Ambisonic mics, derive this signal from eight capsules. Prior literature offers further detail on the geometry of this capsule configuration (Benjamin 2012). Unlike the Ambeo, which offers four XLR analog outputs with a break-out cable, the Spatial Mic transmits digital audio over USB or ADAT, meaning that it applies built-in pre-amplification and analog-to-digital (AD) conversion.

Additionally, we decided to use a 3Dio Free Space Pro II, which is a binaural microphone, as a reference input. The signal from a binaural microphone is head-locked to the position of the microphone during the recording. In other words, it won't respond to 3DoF head tracking and allow the user to rotate their heads in the virtual acoustic space the way an Ambisonic microphone would. Nevertheless, we wanted to compare the performance of the three microphones in a scenario where the listener is facing straight ahead even though they may not be suitable to mix together for head-tracked listening.

5.6.2 *Spatial Layout*

Figure 5.13 shows a top-down layout of the six percussionists, marked with numbers 1 through 6, along with their instruments in Hankinson Hall. The figure also shows the placement of the VR camera, the two Ambisonic mics (Sennheiser Ambeo and Voyage Audio Spatial Mic), the binaural mic (3Dio Free Space Pro II), and the spot mics, which are marked with an "M." While the layout closely follows Chen's instructions in *Hamlet*'s score, it was made more circular than elliptical to ensure each performer was roughly equidistant to the camera and spatial mics.

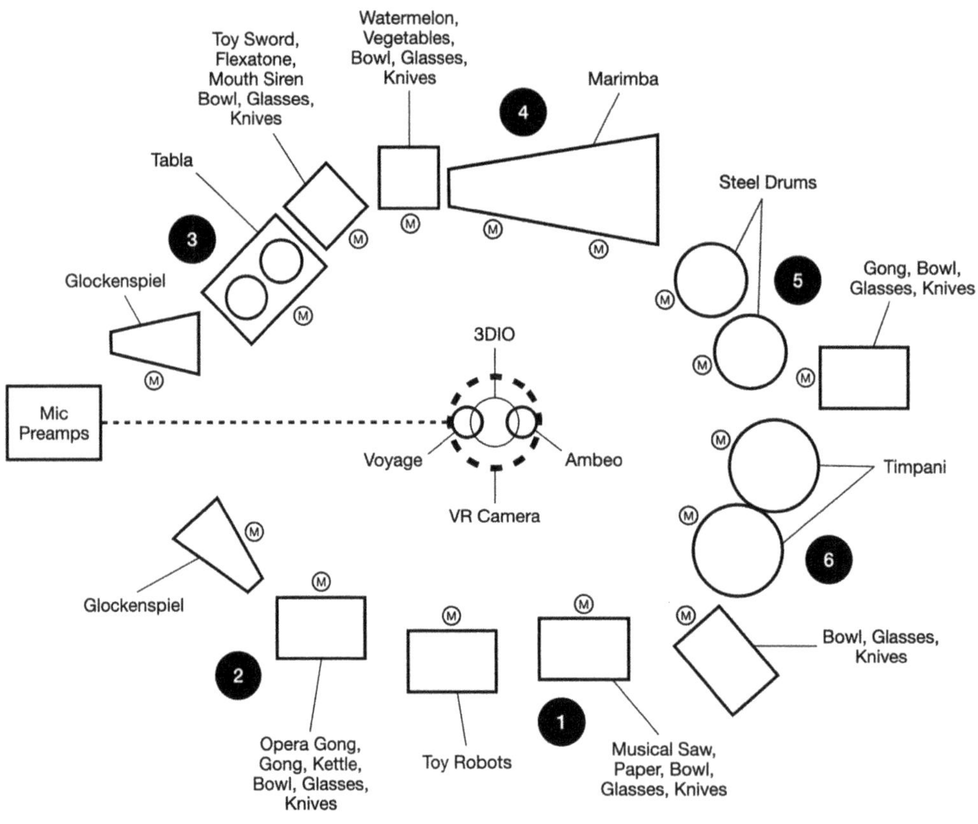

Figure 5.13 A bird's-eye view of the spatial layout for the recording project. This view shows the locations of the performers and instruments, as well as the VR camera and the spatial microphones. The numbers indicate the members of the Percussion Ensemble, whereas the circles labeled "M" correspond to the spot microphones.

5.6.3 Production Setup

On the day of the shoot, we started out by setting up the VR camera near the center of Hankinson Hall, slightly closer to the side with a window to better take advantage of the natural sunlight. Once the camera was set in place, the performers began to set up their instruments as seen in Figure 5.14 with each instrument approximately seven feet away from the camera. During this period, we set up the two Ambisonic mics and the binaural mic in the configuration seen in Figure 5.15.

Once the instruments were set up, we began to place the spot mics at the locations marked with an "M" in Figure 5.13. These consisted of sixteen small-diaphragm condenser mics with cardioid polar patterns. More specifically, we used six Neumann KM 184s, six Sennheiser MKH 40s, and four DPA 4011As. Since the output of the spot mics were intended to augment the ambience captured with the Ambisonic mics, we preferred to have relatively isolated feeds from the spot mics with minimal room effect. Therefore, we chose mics with a directional (i.e., cardioid) polar pattern placed relatively close to the instruments to prioritize the direct sound over the room reflections and minimize bleed from other instruments.

Since this was a live recording involving theatrical elements, the performers couldn't be isolated from each other. Therefore, some degree of bleed was unavoidable between the spot microphones. This is a common aspect of classical music recordings, wherein the performers usually share the same tracking space, whether this is a live room of a recording studio or the stage of a concert hall. Moreover, the bleed between the

Figure 5.14 The Percussion Ensemble's setup for *Hamlet* with the VR camera and spatial mics at the center. The performers and their instruments are organized around the camera in a circle. The window in one corner of Hankinson Hall lets in a considerable amount of natural light.

Figure 5.15 The two Ambisonic mics suspended in an upright position over the VR camera on a boom stand. The binaural mic, which was used for reference, was placed under the camera.

microphones can be a desirable artifact that helps articulate space as long as there are no phase cancellations that may have a destructive effect on the sound. If such cancellations were to occur, they can be mitigated by strategic phase inversions applied during mixing. In our case, the close placement of the spot mics worked in our favor to avoid these types of artifacts.

The audio signals were routed into a DAW on a computer located in a different room through a Dante Network.[13] This was achieved through two Dante Racks that housed Focusrite RedNet 2 preamp and AD converters for up to sixteen channels each. This set up allowed us to dedicate four channels to the Sennheiser Ambeo, eight channels to the Voyage Audio Spatial Mic, two channels to the 3Dio Free Space Pro II, and sixteen channels to various spot mics with two channels to spare. While the Dante Network greatly reduces the need for cabling from the racks to the computer, the routing from the microphones to the racks is fairly traditional and would be the same if we were to use an analog patchbay in a studio or field recording set up.

5.6.4 Video Post-Production

The video post-production involves a process similar to that of the first case study, where we start out by importing the dedicated feeds from each lens of the VR camera into the Insta360 Pro stitching software to composite them into an over-under panoramic video for 3D view. Whereas the first case study involved a seven-minute scene that fit into a single take, a full performance of *Hamlet* was captured in two takes of fifteen minutes that needed to be stitched individually and later combined in Premiere.

Compared to the first case study, this shoot involved more equipment visible to the audience with microphones above and below the camera, as well as a Dante rack for audio routing. While these don't interfere with the visibility of the musicians, they still appear extraneous to the performance and take up visual space. To mitigate this, we captured the performance space one more time at the end of our session once all the instruments and equipment were removed from Hankinson Hall. This gave us a clear picture of the floor, ceiling, and the walls of the hall. From this capture, we cut out the portions where the musicians are located as seen in the second panel in Figure 5.16 and layer the result onto the VR video. In doing so, we block out equipment that could distract the viewer from the performance. We also use an AI fill to cover the small region at the nadir of the camera, where the tripod was visible. The result of the compositing can be seen in the third panel in Figure 5.16, where the camera appears suspended in midair without any equipment nearby. As an auditory counterpart to this process, we also captured the sound of the hall in silence as room tone in case we need to edit out any unintended noises that may have occurred during the rests in the piece.

5.6.5 Audio Post-Production

The audio post-production for the project involves converting the first and second-order Ambisonic recordings to B-format, encoding the spot recordings into a soundfield as point sources in 3D space, and applying general mixing techniques (e.g., leveling, EQ, compression) as needed. As in the previous case study, we use REAPER as our DAW, this time showcasing additional features and different plugins.

The order in which we apply mixing techniques depends on the material we deal with. With Ambisonic recordings, we would want to apply the A-B format conversion before applying any additional effects as these can interfere with the accurate conversion of the original recordings into B-format. After the conversion, we can use multichannel plugins, such as the ones we mentioned in our previous case study, to further process the B-format signal. With spot recordings captured with traditional microphones, we typically apply

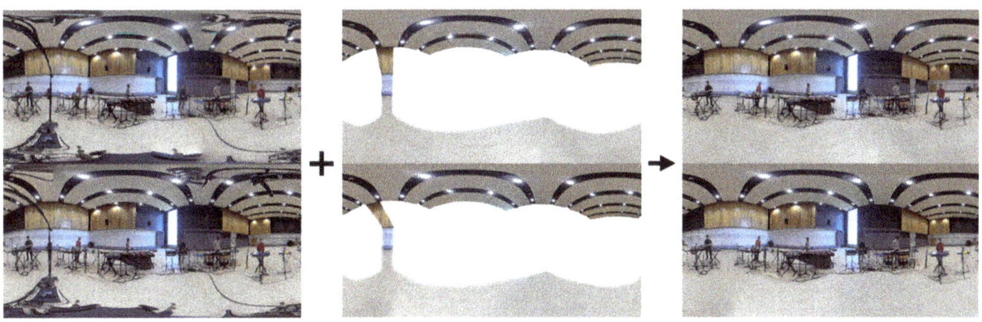

Figure 5.16 Removal of equipment from the VR video using a shot of the performance space after the equipment has been removed. This shot is layered on top of the VR video.

effects like EQ and compression before we encode them into a virtual soundfield. That being said, the result can be further modified with multichannel plugins, especially if we need to process the soundfield as a whole.

In the current project, we were able to work in an acoustically-treated space and capture the performance with a mixture of Ambisonic and spot microphones. In this case, the Ambisonic recordings help us define the space around the spot mics during the mix. However, when an Ambisonic mic isn't available or the recording space lacks desirable acoustic qualities, spot mics can be used to capture a neutral representation of the sound sources at close proximity. These can then be woven into a virtual soundfield, where effects like Ambisonic reverb can be applied to craft a spatial environment around them.

5.6.5.1 Ambisonic Order

As we mentioned in Section 4.3.3, increasing the Ambisonic order encodes the soundfield with greater detail. While the Ambisonic order of recordings made with a soundfield mic depends on the number of channels on the microphone, the Ambisonic order when encoding monophonic or stereophonic sound sources into a virtual soundfield can be arbitrarily defined based on computational resources. With that in mind, one might follow a "higher the better" approach. However, the primary bottleneck in terms of the resolution lies in the currently available delivery platforms. As of this writing, YouTube supports only first-order Ambisonics, so even if a mix is executed in higher-order Ambisonics, the result would need to be downmixed to first-order for YouTube. Meta 360 VR supports a proprietary format called "hybrid higher-order Ambisonics," where eight channels are used to approximate second-order Ambisonics, incorporating "the quality of higher-order ambisonics with fewer channels, ultimately saving on bandwidth."[14] While these platforms may currently limit the Ambisonic order in which a mix is delivered to audiences, we expect these limitations to ease as greater computational resources and internet bandwidth becomes available to consumers. This is one the areas where Ambisonics is particularly powerful as an intermediary format. Once a virtual soundfield is designed, it can easily be encoded into different Ambisonic orders based on the delivery platform. In other words, a mix encoded in first-order Ambisonics for YouTube today could be re-encoded into higher-order Ambisonics if YouTube were to support this format in the future.

While the Ambisonic order of a recording made with a soundfield microphone may be more limited than a soundfield virtually encoded from point sources, soundfields of different orders can be easily mixed together. This is because the channels of a first-order soundfield correspond to the first four channels of a second-order soundfield and the channels of a second order soundfield are the first nine channels of a third-order soundfield, and so on. In other words, as the Ambisonic order increases, the added channels introduce greater detail to the spatial image. Furthermore, there are upscaling plugins that re-encode lower-order Ambisonic soundfields into higher-order ones. With greater incorporation of AI algorithms, we expect these upscaling plugins to perform even better in the near future. In the current project, we will encode our virtual soundfield into third-order Ambisonics and mix it with the first or second-order recordings made with the soundfield microphones. This will give us greater detail during the mixing process and help us demonstrate the channel configurations needed for a hybrid flow involving soundfields of different orders.

Although the output of the DAW project will be a third-order Ambisonic file with sixteen channels, we will need to convert it to binaural audio to monitor it with headphones during the mixing process. In our previous case study, we achieved this with Premiere's Binauralizer plugin. This time we will use the dearVR Ambi Micro plugin shown in Figure 5.17. This is a free plugin that can convert between various Ambisonic orders as well as between Ambisonic and binaural formats based on a selection of non-individualized HRTFs. In our project, we use this on our Master track to convert the sixteen-channel third-order Ambisonic signal coming into the track (shown as "TOA (16CH)" in the plugin) to two-channel binaural audio. In Figure 5.17, the output format of the Ambi Micro plugin is set to "Binaural (YouTube VR)," which emulates how our mix would sound after YouTube converts the Ambisonic signal to binaural. We should reiterate that this plugin must be disabled before exporting the final output, which needs to be in Ambisonic B-format for the head-tracked audio to work as intended in VR.

5.6.5.2 *Processing Soundfield Recordings*

To convert the A-format signals from the two Ambisonic microphones into B-format, we use the conversion plugins provided with the microphones. In our first case study, we routed the four mono tracks from the Ambeo mic to the four channels of the Master track, which had the Ambeo A-B Format Converter plugin. In this project, we apply this conversion before the plugin output is sent to the Master track, where soundfields of different orders will be mixed together. To achieve this, we can route the mono recordings from the individual capsules of the Ambeo to a four-channel aux track that has the converter plugin. Alternatively, we can set up a track folder that contains the four tracks and attach the plugin to the folder. Folder tracks in REAPER are like track groups in other DAWs, wherein the outputs of the group tracks are first sent to a parent track, which itself is routed to the mains. When we create a track folder in REAPER, the folder tracks (i.e., child tracks) are automatically routed to the track folder (i.e., parent track). However, these are setup as stereo sends from the child to the first two channels of the parent by default. In the routing settings of the individual folder tracks, we would need to adjust the "Parent send channels from/to" setting to be a mono send based on the A-format order of the channels (e.g., the third track would need to be routed to the third channel

Figure 5.17 The dearVR Ambi Micro plugin, which can convert between various Ambisonic formats and also output binaural audio for headphone monitoring. Here, the plugin is set up to convert the sixteen-channel Ambisonic B-format signal to binaural based on the standardized HRTF used by YouTube VR.

of the parent). An alternative to this would be unchecking the "Parent send channels from/to" box in the child tracks and adding them as receives in the track folder. This allows all the child track mappings to be viewed in the folder track's routing settings rather than having to browse through the settings for each child track. The latter could become particularly tedious when working with a second or third-order Ambisonic recording, where the channel count increases. Once the folder track for the Ambeo recording is set up, we follow the same process for the Voyage Audio Spatial Mic tracks. We route these to a ten-channel[15] track folder that has the Spatial Mic Converter plugin shown in Figure 5.18. Note the similarities between this A-B format converter plugin and the one for Ambeo, which we previously showed in Figure 5.1.

Ultimately, we will use only one of the two soundfield recordings in the final mix to avoid phase issues that may occur due to the near-coincident placement of the two Ambisonic microphones. Nevertheless, both recordings are available to the readers in the project files provided at vraudio.org under Chapter 5. We recommend switching between the two to compare the sense of space they each convey. Bear in mind that these are not two identical microphones simply with different Ambisonic orders. There are numerous factors that contribute to the quality of a microphone, such as sensitivity, frequency

Figure 5.18 Voyage Audio Spatial Mic Converter plugin. This plugin converts the output of the Spatial Mic to various formats including first and second-order Ambisonics, mid-side, and surround. It offers controls similar to those that we saw in the Ambeo A-B format converter and dearVR Ambi Micro.

response, equivalent noise level, signal-to-noise ratio, and total harmonic distortion. If you inspect the specification sheets of these two Ambisonic mics, you will notice that they differ in all of those areas. However, you can still compare the spaces that you perceive in these recordings. Alternate between the first and second-order Ambisonic recordings by soloing one track folder or the other. Do you hear a change in the placement of the instruments or the distance between them? How about their distance to you? Do they feel closer or farther away? Do you perceive a change in the overall height, depth, or width of the space? Besides their spatial features, do you notice other differences in quality between these microphones? Is one noisier than the other? Can you hear a difference in their frequency response? You can also focus on how the two interact with the close mics. Mute one of the Ambisonic mic track folders to hear the close mics in the context of the other Ambisonic recording.

5.6.5.3 Processing Spot Recordings

Once the channel routing for the Ambisonic mic recordings are set up, we move on to the spot mics. When mixing music, we often apply effects like EQ and compression to process the timbre and dynamics of a track. Certain genres like pop and rock can benefit from heavy processing that significantly alters the material recorded in the studio. However, classical music production often involves subtle modifications that do not substantially interfere with the authenticity and liveness of a performance. Nevertheless, we may still want to apply filtering and dynamics processing to bring out certain musical qualities and balance instruments in the mix. Despite its experimental qualities, the piece we recorded in this project can be considered a part of the classical idiom. Therefore, while we decided to apply some EQ and compression, these are kept at a subtle level to maintain the natural qualities of the performance. As we mentioned earlier, these are applied before the mono recordings from the spot mics are encoded into an Ambisonic soundfield. Figure 5.19 shows the REAPER project where we carry out the mix. Here, we see the spot recordings grouped under a track folder titled "3rd-order Soundfield," which we will use to encode the recordings as point sources.

For the encoding, we can use a plugin like the O3A Panner on the individual spot mic tracks. However, to demonstrate a different workflow than the one we used in the first case study, we instead use the IEM MultiEncoder plugin shown in Figure 5.20. This is a free plugin that can encode multiple monophonic signals into a soundfield of up to seventh-order Ambisonics. This plugin can be attached to a track folder that contains the spot mic tracks or a dedicated track which the spot mic tracks can be routed to. The Master controls in this plugin can be used for rotating a soundfield with all the point sources in it. Since we are aiming for third-order Ambisonics, the channel count for this track should be set to sixteen based on the $(N+1)^2$ formula, where N represents the Ambisonic order. Coincidentally, we had sixteen spot mics in our recording. If we happened to have more, we would increase the channel count to accommodate the additional inputs, but the output would still take up only the first sixteen channels.

To retrieve the azimuth and elevation values for our instruments, we can still use the image loading feature of the O3A Panner (Large) or a similar plugin for visual reference. We can then use these values in the IEM MultiEncoder plugin to position the individual tracks as point sources based on their location in the visual panorama. The VR video and the Ambisonic recording can serve as spatial references even if an encoder plugin does not

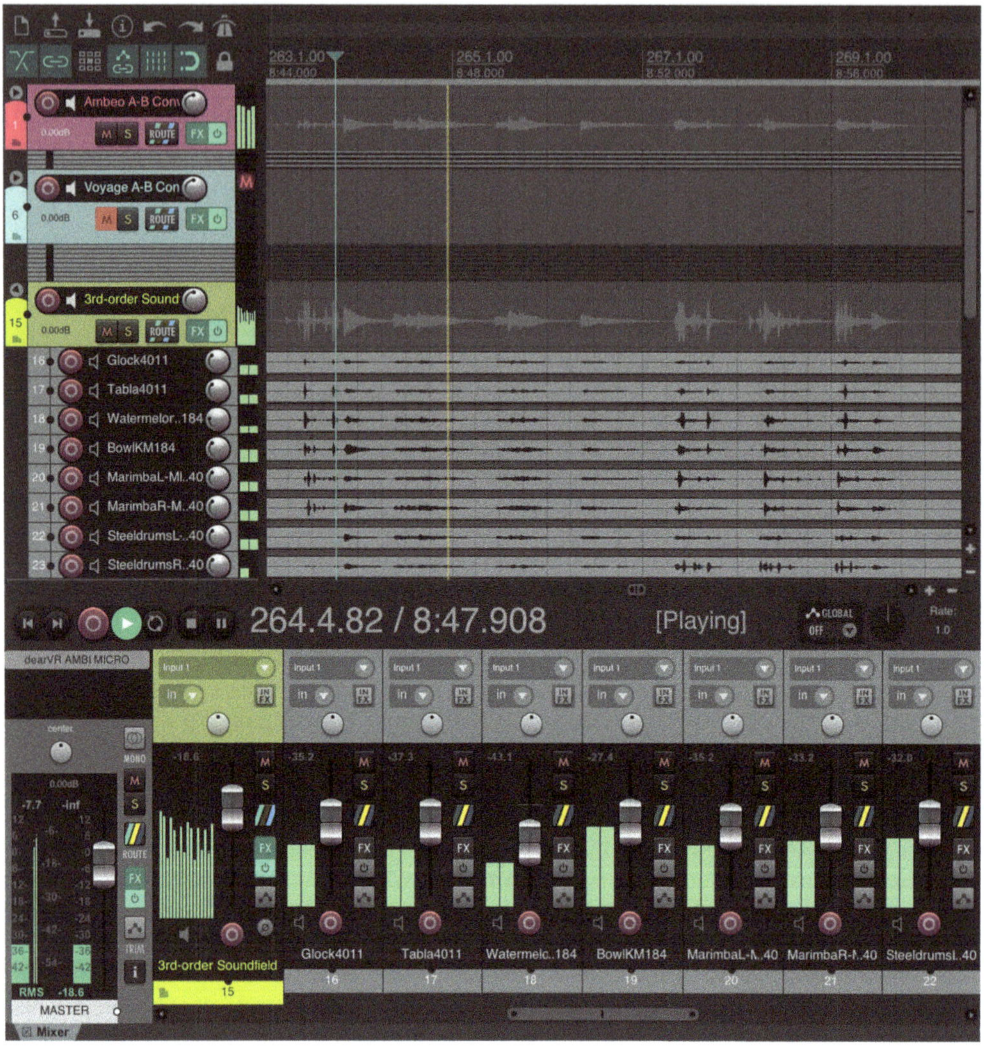

Figure 5.19 REAPER session for the project. The Ambeo and Voyage Audio track folders are collapsed to take up less vertical space. In this image, we are monitoring the mix of the Ambeo recording and the third-order Ambisonic soundfield created from the spot mic recordings. We use the dearVR Ambi Micro on the Master track to convert the third-order output to binaural for headphone monitoring during the mixing process.

offer image-loading functionality. It is worth noting that the virtual positioning of these point sources in an Ambisonic soundfield is not an exact science. While there is a case to be made for realism when mixing these sounds, there are other factors, such as spatial separation and clarity, that can prompt the designer to rely on their ears and deviate slightly from an accurate simulation of a soundfield. Nevertheless, we must ensure that the recorded soundfield and the one encoded from point recordings blend together

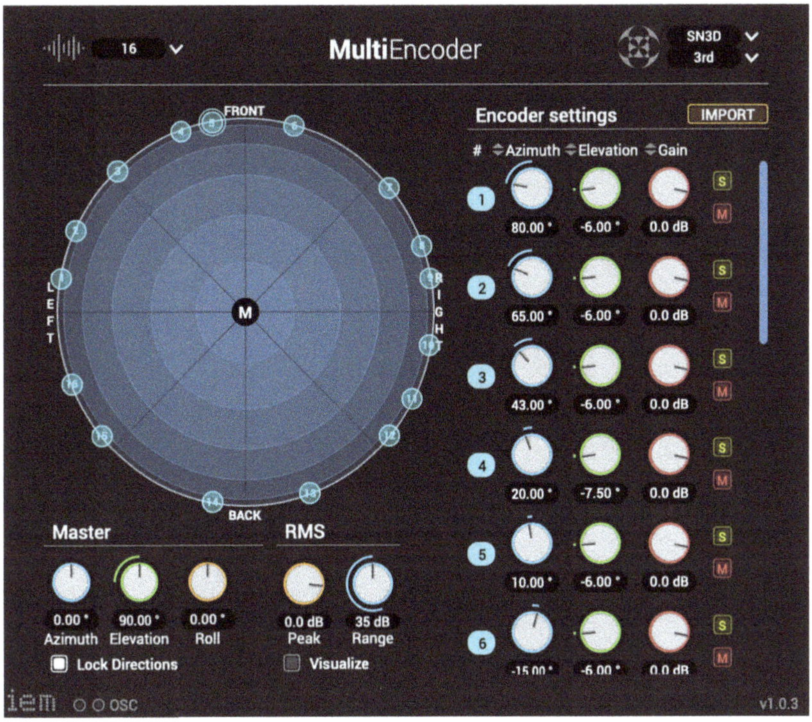

Figure 5.20 IEM MultiEncoder plugin. The spot mic tracks are routed to the consecutive inputs of a track folder that has this plugin. The monophonic inputs are positioned using the azimuth and elevation parameters dedicated to each input. The output of the plugin is a third-order Ambisonic soundfield.

cohesively. For instance, we can mute and unmute the individual point sources to inspect how they fit within the panorama established by the Ambisonic recording. Once we are satisfied with the placement of the point sources, we can adjust the level of the Ambisonic recording to balance the amount of reverberant sound in the mix.

5.6.6 *Exporting*

Since we carry out the audio mix in a DAW, we don't have to do much audio processing in Premiere before we export the film. If the audio format for the intended delivery platform is first-order Ambisonics, we may use a plugin like the dearVR Ambi Micro on the Master track in REAPER to downscale the third-order Ambisonic mix as we render it. We would then bring this audio file into an adaptive audio track in Premiere and align it with the video using the camera's audio recording as a reference. We would then set up the export format based on the delivery platform recommendations before we export our media file. If the delivery platform supports higher-order Ambisonics, we may need to export the audio and video separately and combine them, as Premiere imposes a limit on the number of audio channels based on the video format. To this end, we can use a multiplexing (muxing) software to combine the VR video and the Ambisonic audio.

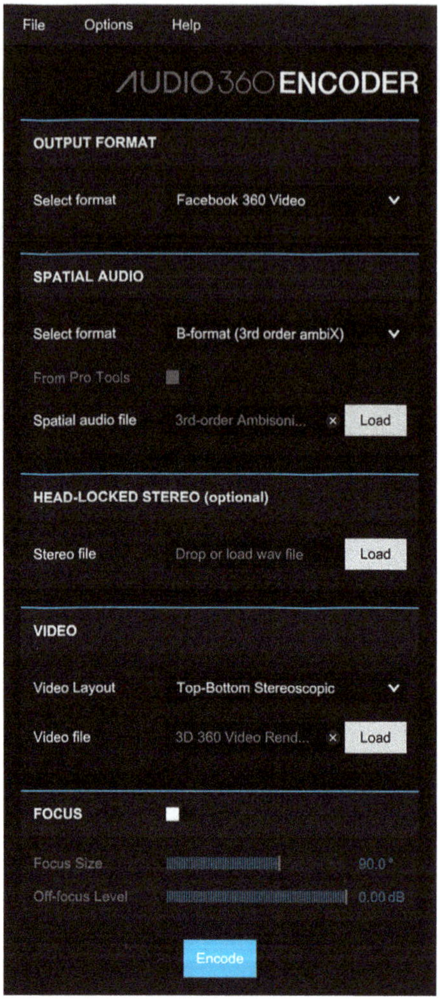

Figure 5.21 Encoder tool provided by Meta, allowing the user to combine Ambisonic and head-locked audio tracks to VR video.

In Figure 5.21, we see an encoder software offered by Meta for combining these media files based on presets for various delivery platforms. Here, the user can add an Ambisonic track, an optional head-locked stereo track, and the VR video to combine them into a single media file. Other muxing software, such as FFMPEG, can be used for this purpose as well.

In the next chapter, we will move onto applications of VR audio in interactive media, with a particular focus on games, where the user will be able to not only experience the virtual environment with six degrees of freedom but also engage with the elements in it. These affordances will bring about a whole new set of techniques and considerations, which we will explore through case studies.

Notes

1 Yes, this is a Monty Python reference.
2 A shotgun microphone is a specialty microphone with a highly directional polar pattern. The directivity of the microphone allows the boom operator to isolate a sound source while rejecting other sounds that are nearby.
3 In applications where the spatial accuracy of point sources is not critical, an Ambisonic microphone alone can afford a sufficient sense of space. Furthermore, higher-order Ambisonic microphones, such as the Core Audio Octamic, Zylia ZM-1, Voyage Audio Spatial Mic, and Eigenmike em64, are becoming readily available. With higher-density capsule arrays, it is possible to capture soundfields with greater spatial detail.
4 https://web.archive.org/web/20230927172929/https://en-us.sennheiser.com/ambeo-abconverter.
5 This often falls onto the production designer, who manages the visual style of any given scene by choosing locations, designing or sourcing props, and decorating the scene. In productions with smaller budgets, the director, cinematographer, and production designer might be one and the same person.
6 https://www.nfb.ca/interactive/notaceremony/.
7 https://afterlife-vr.com.
8 https://www.sexradicalmovie.com.
9 https://www.reaper.fm/.
10 REAPER has historically offered a trial period that can be extended indefinitely, making it an invaluable tool for enthusiasts and students. If you happen to benefit from REAPER in a professional context, it is recommended to purchase a license to support a sustainable future for this tool.
11 http://www.blueripplesound.com/o3a_core.
12 https://plugins.iem.at.
13 Dante is an Ethernet-based audiovisual network protocol, which allows low-latency uncompressed audio to be transmitted between nodes on the network, functioning as a digital patchbay.
14 https://engineering.fb.com/2017/02/22/virtual-reality/spatial-audio-bringing-realistic-sound-to-360-video/. Accessed on December 5, 2025.
15 Although this plugin has only nine outputs for second-order Ambisonics, REAPER channels can only be set to multiples of two.

Chapter 6

Sound in VR Games

Crafting Audio for Interactive VR

Sound plays an inextricable role in video games: dialogue helps the game story move forward, sound effects signify events or objects in the game world, and music sets the mood for the gaming experience. Even before we consider the overlap between VR technology and gaming, sound has been deemed critical for creating an immersive experience in games (Gallacher 2013; Broderick, Duggan, and Redfern 2018). Sound design for video games is a well-established field that has been extensively covered in prior literature (for a few examples, see Marks 2017, Horowitz and Looney 2014). Some of the concepts and techniques employed in traditional game audio are applicable to VR development. However, VR games also require a different set of considerations for audio design as emerging tools and techniques for implementing immersive audio in game engines bring about unique workflows and creative possibilities. Moreover, audio in VR games assumes new narrative functions beyond its conventional role in storytelling. It can guide the user and help them orient themselves in VR, it can reinforce or subvert the user's senses of space and agency, and it can articulate distinctions between the diegetic and non-diegetic elements of a game.

In this chapter, we build on the discussion of spatial audio recording and production for VR film in Chapter 5 and turn our focus to the computational modeling of immersive sonic environments for interactive VR experiences, including games. We examine fundamental aspects of spatial audio design, such as sound object types, source directivity, and room effects, within the context of VR games. We then look at audio implementations in modern game engines, focusing on the spatial and dynamic relationships between sounds and game objects. In doing so, we build upon traditional game audio workflows and identify emerging guidelines for VR sound design. Furthermore, we investigate the application of procedural audio in VR games and highlight its increasingly critical role in immersive experiences. Much like in Chapter 5, our approach to these topics incorporates practical case studies, in which we discuss not only the technical but also aesthetic considerations that guided our designs. More specifically, we explore the implementation of two VR music games using two of the leading game engines today.

6.1 Games Versus Films in VR

While there are similarities between the concepts and techniques involved in sound design for VR films and VR games, the latter brings about distinct considerations due to the inherently interactive nature of gaming. For instance, the VR film projects that we discussed in Chapter 5 offer the user three degrees of freedom (3DoF): pitch, yaw, and roll.

DOI: 10.4324/9781003048022-6

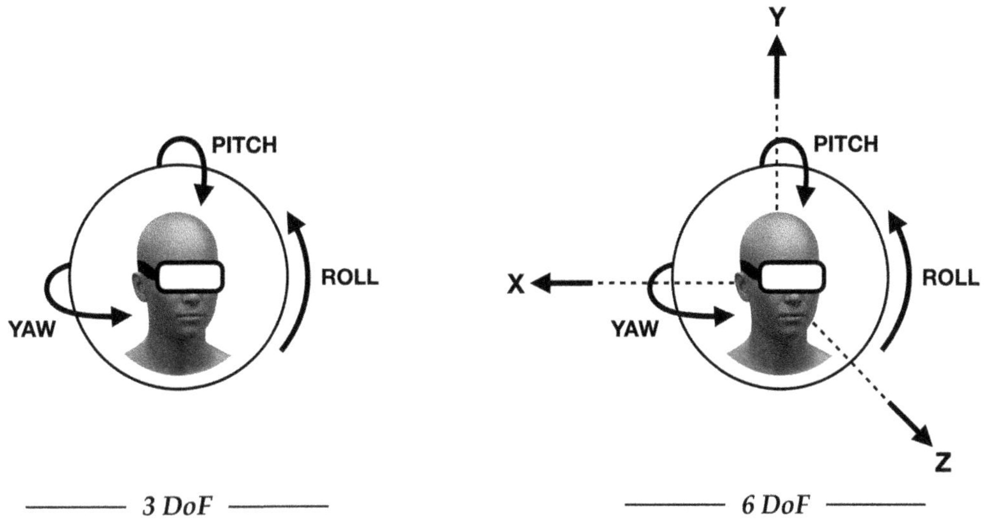

Figure 6.1 On the left: 3 degrees of freedom (DoF), where the user can rotate their head freely to view VR space, wherein their position is locked. On the right: 6DoF, where the user can not only rotate their head to view VR space but also move through it in the X, Y, and Z coordinates.

Although a user can look around freely in a 3DoF VR film, they cannot move through virtual space as their position is locked to where the camera was placed during production. In VR games, however, the user can not only look around but also navigate the virtual environment with 6DoF. Figure 6.1 visualizes the distinction between 3DoF and 6DoF VR experiences, which we previously touched upon in Section 4.2.3. In 6DoF, the rotational degrees of freedom that pertain to pitch, yaw, and roll are augmented by positional degrees of freedom, more specifically the Cartesian coordinates of X, Y, and Z. The user can traverse these dimensions either through virtual locomotion or room-scale movement in physical space.

Additionally, games and gamified experiences often incorporate interactive objects, navigable environments, and non-player characters (NPCs). Therefore, VR games can impart a greater sense of agency to the user. Having said that, as practices that are native to the immersive medium (rather than those that are adapted to it) continue to emerge, we expect the distinctions between films and games (or fixed and interactive media) to ultimately fade away. Particularly, with greater computational resources becoming available for consumer applications and the rise of generative AI models that can interpret and synthesize media content in real time, VR games will achieve a higher degree of photorealism, while VR films will become increasingly interactive. Keeping this possible hybrid future in mind, let's explore some of the distinct concepts and techniques relevant to VR games.

6.2 Producing Sound for VR Games

Sound for games and other interactive media can be created in several ways, some of which will be familiar from our conversation on VR film sound in the previous chapter.

For example, game sounds can be recorded from real-life sources using microphones. When recording a sound for a game, many of the techniques and considerations applicable to studio recording come into play. These include the proper choice of microphone for the recording task at hand, the correct placement of the microphone, the balance and clarity of the sound source versus any ambient or extraneous sounds that may be present in the recording environment, the mitigation of noise that could occur in the signal path due to faulty equipment or electrical interference, and proper gain staging.

If a game element represents a real-life object (e.g., a clock in the game environment), its sound can be recorded from the said object either in a studio or on location. Furthermore, recordings of various objects can be amalgamated into more elaborate sound effects. For instance, recordings of broken glass and struck metal can be combined with a low frequency impulse and filtered noise to emulate the sound of an explosion. Recorded sounds can also be manipulated to create sound effects that don't have a real-life referent. For example, recordings of animal sounds are often pitch-shifted down to produce monster voices in fantasy games.

Game sounds can also be generated from scratch using hardware and software synthesizers and audio programming tools. Audio synthesis can emulate real world sounds in the style of Foley, but it can also be leveraged to design abstract or otherworldly sounds. Entirely new sounds can be crafted with techniques like subtractive, frequency modulation, and granular synthesis. These sounds can be further manipulated with time- or frequency-domain signal processing techniques, giving them greater realism or making them even more alien.

The case studies presented later in this chapter incorporate a combination of these methods for game sound design. For example, we demonstrate our use of location recordings mixed with synthesized sounds to establish an ambient texture for a scene. This mixture serves a background function that evokes real-life imagery through representational sounds while at the same time setting the mood for the scene with more abstract sounds. These are then overlaid with recorded and synthesized sounds that originate from player interactions. By integrating various sound reproduction, generation, and processing techniques, we can craft rich, immersive audio experiences.

6.3 Procedural Audio

A technique that finds increasing use in games is procedural audio, which involves the real-time generation and processing of sounds based on gameplay parameters. The use of this technique is directly correlated with the increase in the computational resources available to developers. One of the defining qualities of a modern computer is its ability to perform complex logical tasks quickly, repeatedly, and with great precision. We can feed it a convoluted set of instructions and expect it to carry them out with consistent results at a pace that is far beyond human ability. Instead of telling a computer exactly what it needs to do, we can also provide it with algorithms that enable the computer to make decisions or predictions based on a range of variables including user input or external data sets. This allows us to design computational systems that are not only reactive but also generative. For instance, a computer can employ genetic algorithms that simulate the behavior of an organism's cell structures to predict its evolution. These types of processes are also used in art to create audio and visual materials. A common example of algorithmic art involves the use of fractals to generate complex visual patterns.

The past few decades have seen major advances in computational technology ranging from faster processing units to larger memory components. These advances enable computers to perform complex algorithmic tasks in real time, opening up new creative possibilities. This, for instance, allows video games to incorporate procedural audio and visual events that take form during runtime without relying on a predetermined order or timing (Farnell 2014). As a result, an increasing number of games today follow non-linear trajectories that are unique to each individual player.

Game sound researcher Karen Collins argues that most modern games can be considered to have procedural soundtracks by way of incorporating musical events that get triggered and evolve in real-time based on player actions (2009). The procedure underlying this evolution can be as simple as detecting when the player enters a certain area in the game map and triggering a pre-recorded piece of music that complements the story unfolding in this particular area. However, procedural audio can also be more elaborate and generative. The non-diegetic music that accompanies the in-game sound effects can be composed and synthesized in real time to create musical structures that are unique to each gameplay. The sound effects themselves can be synthesized in real time based on physical models governed by user interactions and the features of the game environment. This type of procedural audio generation allows each game interaction to produce a distinct sound, similar to what we would expect from our interactions with the real world.

A simple example of this involves the design of footstep sounds for games. Throughout the history of video games, footstep sounds have evolved alongside advances in audio technology. Therefore, these sounds illustrate how far we have come in game audio. Especially in first-person and third-person role-playing games, footstep sounds play a critical role in reflecting the player character's locomotion through the game environment. They can indicate not only movement speed, but also the surface material of the environment that the character is traversing. Furthermore, footstep sounds can be used as a game mechanic to signal how stealthily a player is moving (Ekman 2005).

Initial implementations of footstep sounds in games were relatively rudimentary. In early arcade machines and consoles, audio cues were delivered through dedicated chips that were purpose-built to play primitive sounds such as basic waves and noise bursts. Footsteps had a fixed sound that did not bear a great resemblance to the sound of an actual footstep and was only meant to represent the transient and repetitive nature of walking. Even when audio playback became available in games, only a low-resolution recording of a single footstep sound could be used due to memory constraints of these early systems. This resulted in sounds that were more clearly indicative of walking, but lacked the natural variation between footsteps that one would hear in real life. Moreover, a single sample can become pretty grating when played back repeatedly with each step. Therefore, as advances in computational technology allowed more memory to be allocated to audio files in games, designers began to use two or more samples for the same action. This enabled each step taken in a game to trigger one from a selection of footstep sounds to avoid repeating the same sample and achieve a more natural auditory representation of walking. Then came the era of real-time digital signal processing in games, where, instead of merely picking from a set of samples, the selected sample could be modified with changes in pitch, dynamics, and filtering to introduce even more variation to each step.

Today, we can design fully procedural footstep sounds that are processed or synthesized in real-time based on various game features such as the character's footwear, walking pace, and the surface they are traversing. This allows for a level of complexity that approximates the acoustic properties of real-life footsteps. As a result, footsteps in modern games sound much more natural and reactive to the user's actions. In a sense, footstep sounds epitomize the evolution of game audio, going from one-shot noise bursts to complex sounds generated procedurally during gameplay.

Given the elevated sense of agency and interactivity that VR affords its users, the use of procedural audio in a VR game can have a profound effect on the plausibility illusion, namely a sense that the events in VR are actually happening (Slater et al. 2022). With procedural audio, each interaction on the user's part can be made to have a unique auditory response, more closely emulating our interactions with the physical world. The degrees of freedom added to user interaction via room-scale movement, hand-held controllers, and the tracking of the user's body, hands, and eyes offer a wide range of data points to be fed into a procedural audio system in real time. Paired with the spatial audio capabilities of VR, procedural audio can help create rich and dynamic sonic structures in VR games. In Sections 6.7 and 6.9, we will explore the design and implementation of two VR music games that incorporate fully procedural sounds in modern game engines.

As we briefly touched upon earlier, generative AI is finding increasing use in media production for video games, including those for VR. The applications of AI in video games range from procedural generation of models and environments to determining the behavior and features of NPCs and game objects. With the rapid advent of large-language models in the 2020s paired with AI-based voice synthesizers, games have already begun to utilize real-time dialogue generation, where NPCs can respond to the player based on the circumstances of each gameplay. Furthermore, recent research has showcased the use of AI for real-time generation of video game music (Bossalini, Raffe, and Andres Garcia 2020) and sound effects (Marrinan et al. 2024). In the coming years, we expect this type of research to make its way into game audio workflows. This will allow procedural audio to be further integrated with the overall design of a game and be governed by the unique actions of each player.

6.4 Audio Middleware

A common approach to game audio implementation is the use of audio middleware. These are third-party applications that work alongside a game engine to manage audio events and associated assets. Audio middleware is particularly useful for compartmentalizing the game and audio design aspects of a project, allowing audio processes to be handled separately. This can be beneficial for larger game projects, where dedicated teams work on different aspects of a game like environment design, lighting, modeling, and sound. Moreover, audio middleware inherits features from audio production software such as editors and DAWs, making them familiar for audio specialists to use.

The two most popular audio middleware applications are Wwise and FMOD. Both tools allow audio assets to be organized into banks and attached to events that are triggered by the game. They also offer effects and plugins for modifying audio assets in terms of pitch, dynamics, filtering, spatialization, and more. Despite offering similar functionality, Wwise and FMOD take different approaches to their user interfaces. As shown on the left side of Figure 6.2, Wwise offers mixer hierarchies that encapsulate work units. These units can involve Sound SFX, which are audio assets, or switch containers that can select from a selection of Sound SFX. Complex audio events can be crafted using these hierarchies. On the other hand, the FMOD user interface seen in Figure 6.3 prioritizes an

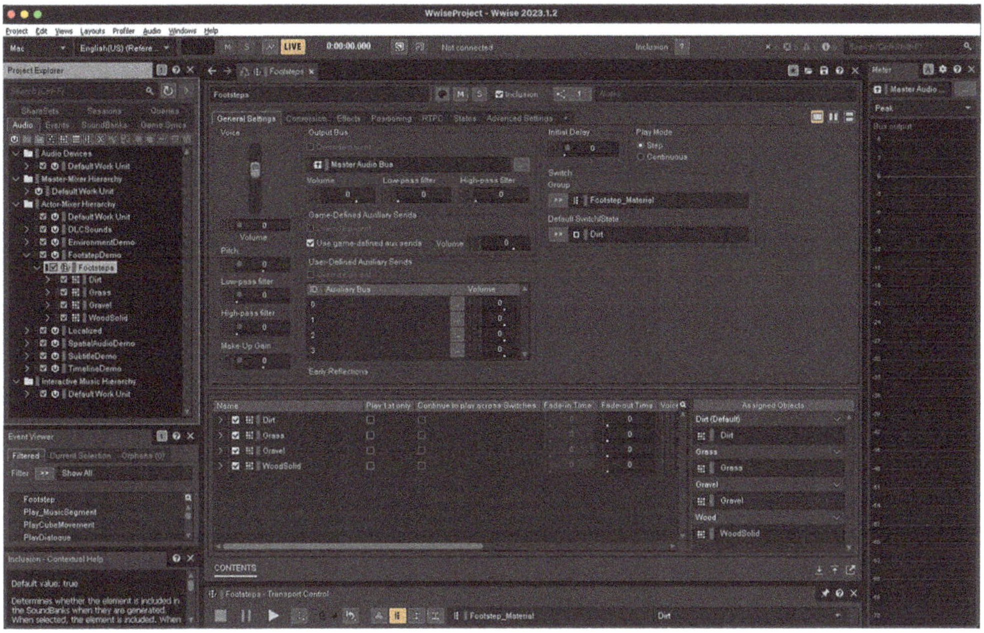

Figure 6.2 Wwise Interface displaying a switch container that can choose from a selection of different footstep sounds. The container allows audio parameters of these sounds to be modified globally.

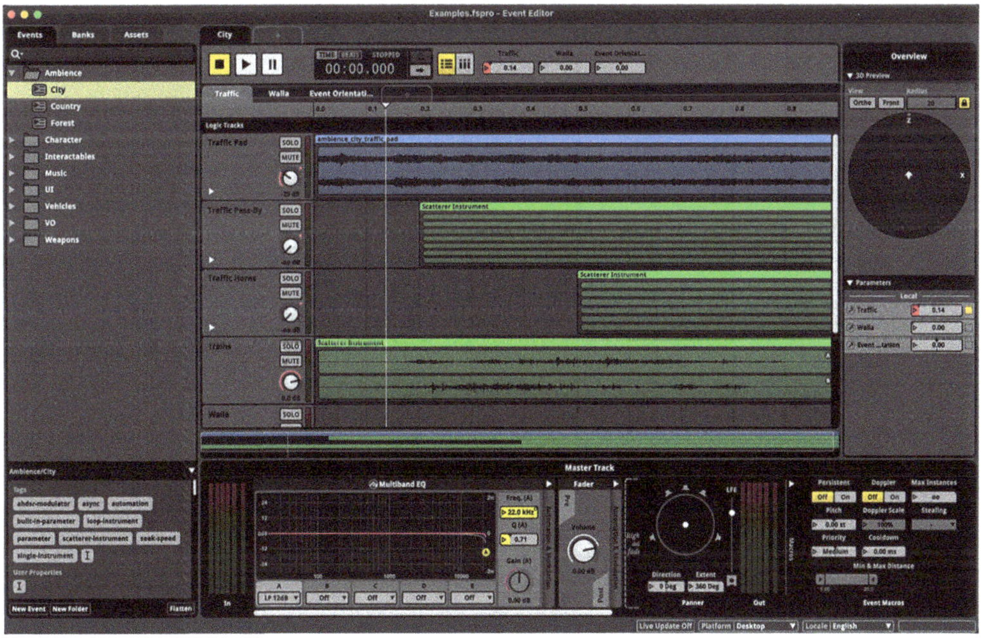

Figure 6.3 FMOD Studio interface with an event timeline in the middle and audio effects applied to this event at the bottom. The panel on the left allows the user to switch between different events, banks, and assets.

event timeline. While FMOD similarly utilizes banks and events, the audio assets are organized within this timeline, which is similar to how files are organized in a DAW. This similarity can make FMOD more approachable to audio specialists who may lack experience in game audio. In fact, the FMOD interface is reminiscent of the popular DAW Ableton Live.

There is extensive literature and numerous online tutorials on the use of audio middleware for game audio design (for a few examples, you can refer to Robinson 2019; Kastbauer 2014; Beauchemin 2016). Since this book is aimed at independent developers, artists, and students who are getting started with VR audio design, and in the interest of expanding the available literature in this domain, this chapter will focus on in-engine workflows. In other words, rather than using audio middleware, we will adopt an approach to VR sound design that is fully implemented within the game engine. Nevertheless, the concepts and techniques that we discuss in this context will also be applicable to audio middleware workflows.

6.5 Spatialization

An area of game audio that gains greater emphasis in VR is sound spatialization. In Chapters 3 and 4, we examined our perception of sound in space and the implementation of spatial audio through computational methods. Although sound spatialization already plays a significant role in traditional video games, it requires further attention in VR due to the first-person and dynamic perspective of the user. While a traditional game could get away with a less elaborate spatial audio design, the same would interfere with the player's sense of immersion and presence in a VR application. Therefore, it is critical for a VR audio designer to have a solid understanding of spatial audio concepts and how these apply to VR experiences. Let's begin by exploring some of these concepts using the web-based immersive audio design tool Inviso accessed at inviso.cc. The Inviso scenes referenced throughout this section can be found at vraudio.org under Chapter 6.

6.5.1 Listener Node

In spatial audio, sound sources are localized in relation to a listener node situated within a computationally defined virtual space. In most spatial audio tools, the listener node is a stationary reference point, typically positioned at the center of the scene. The user perceives the sound sources added to the virtual space from the vantage point of this listener node. Figure 6.4. shows an example of this in the Dolby Atmos Renderer suite. In this figure, the left panel shows the Dolby Atmos Music Panner VST plugin for DAWs. The node labeled 11 allows us to move the incoming audio around a listener node, which is represented with an overhead view of a human body.[1] This control is mirrored in the Dolby Atmos Renderer with a point source situated in a theatre, as seen in the bottom right corner of Figure 6.4. When the panner plugin is attached to other channels in the DAW, these appear as additional nodes in the Renderer. The movement of the sounds in the composed environment is executed in reference to the stationary listener node at the center.

In VR, the listener node itself is a dynamic point in space. As the user moves their avatar through virtual space, or as they move physically within a room-scale VR environment, the listener node moves with them. As a result, the auditory elements in the virtual

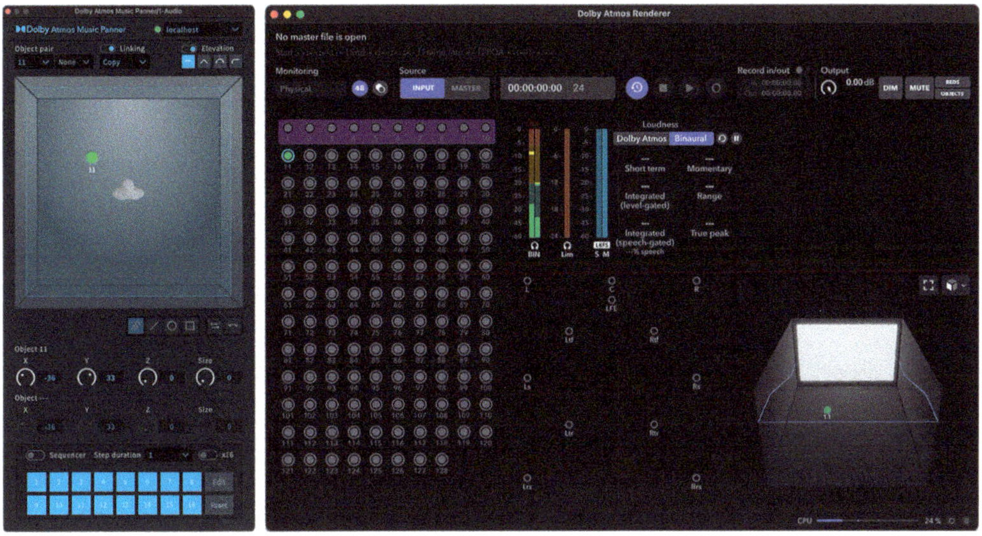

Figure 6.4 Screenshots from the Dolby Atmos Renderer Suite. On the left is the Dolby Atmos Music Panner plugin. On the right is the Dolby Atmos Renderer.

environment are spatialized relative to the listener node's position at any given time. Most commonly, the listener node is attached to the virtual camera that feeds a view of the VR environment to the user's head-mounted display. The position and orientation of the listener node follows the 6DoF tracking of the head-mounted display. A simulation of this is shown in Figure 6.5 using Inviso. Here, we see the listener node moving across a virtual soundscape on a predetermined motion trajectory. This results in a dynamic binaural rendering of the soundscape based on the locomotion of the listener node through 3D space.

In Section 3.2, we described several cues that help us discern the location of a sound source. These included interaural time and level differences as well as pinna and body masking effects. When a listener node moves through virtual space, these cues are calculated and applied to the sounds in that space based on the spatial relationship between the listener and the sound sources. In addition to hearing sounds localized in 3D space, we may hear, for instance, a slight muffling effect if a sound happens to be behind the listener. In combination, these cues help simulate the perception of sound in space. However, there are other properties besides the position of a sound that govern its propagation through virtual space, such as directivity, attenuation, and occlusion.

6.5.2 *Source Directivity*

We can categorize spatial sound sources under three groups based on their directivity: omnidirectional sounds (point sources), unidirectional sounds (sound cones), and nondirectional sounds. Through a combination of sounds with different directivity patterns, we can create spatially rich environments encompassing sound elements that serve a variety of functions. These can range from sounds that are intended to draw the player's attention or move the game forward to those that establish an auditory context or mood for the game environment.

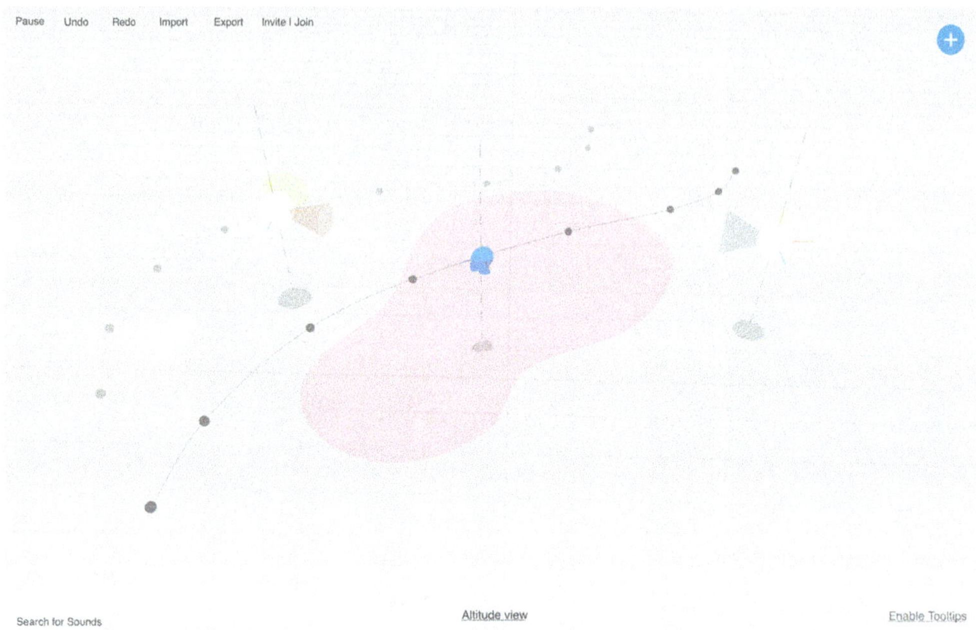

Figure 6.5 An Inviso scene with the listener node (represented with a 3D head model) moving across the scene on a predetermined motion trajectory. You can test this scene in Inviso by importing Inviso_Introduction.zip, which can be found at vraudio.org under Chapter 6.

Omnidirectional sounds, also known as point sources, uniformly propagate in all directions from a point in space. Based on the positioning and distance of the sound source relative to the listener node, an omnidirectional sound is heard as coming from a particular point in the scene at a certain intensity. As either the listener node or the omnidirectional sound moves in the scene, the localization of the sound is updated to maintain its spatial relationship to the user. If an omnidirectional sound is within hearing distance without any occlusions between its source and the listener, it will be audible regardless of its orientation.

In Figure 6.6, we see a listener node on a predetermined trajectory, moving through a space populated with two omnidirectional sound sources shown as spheres. Whereas one of these sources is stationary, the other one moves on a trajectory of its own. The sounds from these sources propagate through 3D space in a spherical pattern (i.e., uniformly in all directions). You can test the scene shown in Figure 6.6 by downloading Inviso_Omnidirectional.zip from vraudio.org and importing it into Inviso at inviso.cc. Try changing the positions and motion trajectories of the objects and the listener node to see how these affect the localization and audability of omnidirectional sound sources.

Unlike an omnidirectional source that emits sound in all directions, **unidirectional sounds** have a limited field of projection. This type of sound models the directivity patterns of real-life acoustic phenomena. For instance, when we speak, our vocal tract and mouth act as an acoustic waveguide that pushes our voice in the direction we face. That is why when we want someone to hear us more clearly, we turn towards them. While

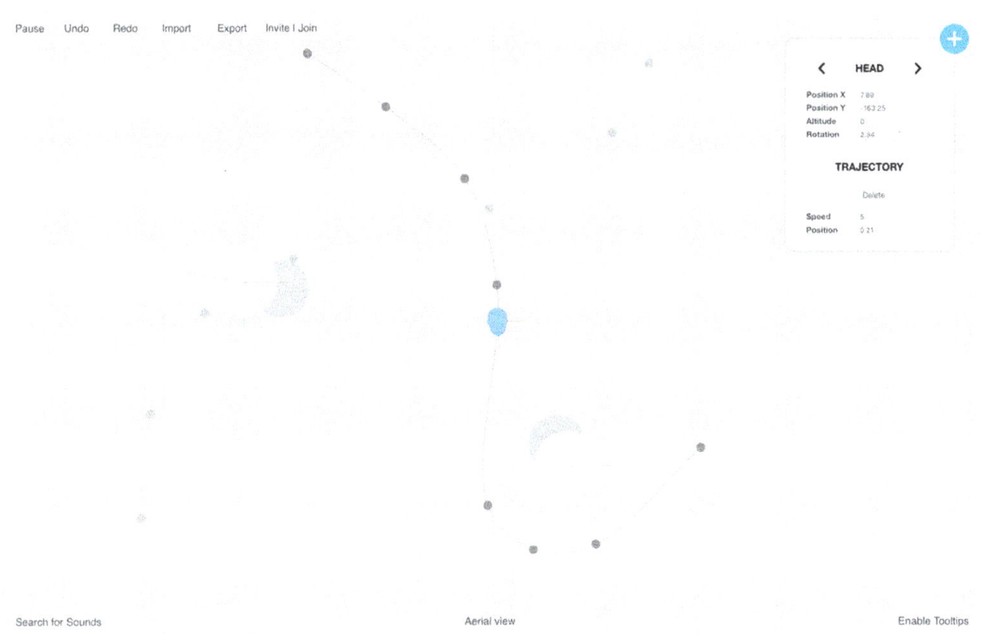

Figure 6.6 An Inviso scene with two omnidirectional sound sources. One of these sound sources moves on a predetermined motion trajectory. The listener node is moving between these sources on a predetermined trajectory of its own. You can test this scene in Inviso by importing Inviso_Omnidirectional.zip, which can be found at vraudio.org under Chapter 6.

someone behind us might still hear us, our voice will be less clear since some of the higher frequencies, which are crucial to speech intelligibility, will have diminished and the overall intensity of our voice will have been reduced. As we discussed in Section 3.5, we encounter similar acoustic principles in musical instruments as well. In addition to the body of an instrument that might project sound omnidirectionally, features like the sound hole in a stringed instrument or the bell on a brass instrument will focus the acoustic output of the instrument in a particular direction.

To achieve this kind of directivity in spatial audio, we use sound cones. When we want our voice to travel a farther distance in real life, we may cup our hands around our mouth or use a cone-shaped waveguide like a megaphone. This helps to concentrate our voice in a certain direction. This is also why spatial audio applications often use cones as a visual metaphor for sounds with limited directivity. Figure 6.7 shows an example of this in Inviso, with a sound object that has three cones attached to it. Note that sound objects in Inviso can be populated with an arbitrary number of cones to create complex directivity characteristics.

When we use a cone to bring directivity to a spatial audio source, we actually utilize two cones nested together: An inner cone which projects the sound source at unity gain to represent the direct portion of the sound propagation, and an outer cone which has a lower gain and, in some cases, filtered sound to represent the indirect portion of the sound propagation. This way, if the listener node is within the field of projection of the inner cone, the user hears a direct sound (like hearing someone talk in our direction).

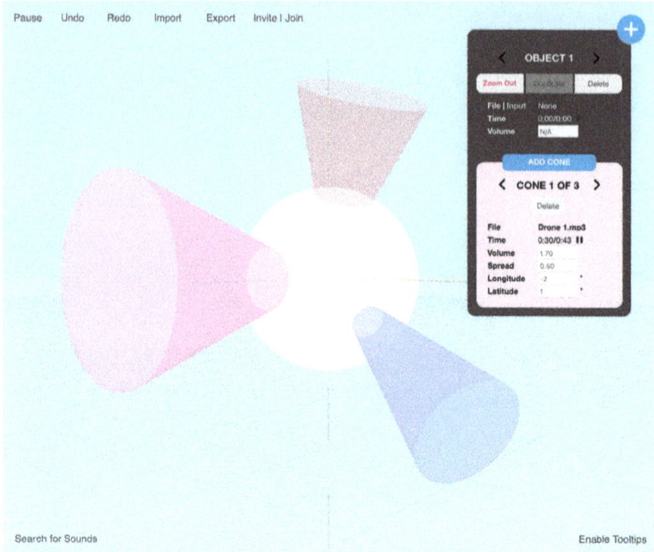

Figure 6.7 An Inviso scene, where we are currently zoomed into a sound object with three cones. The parameter window allows the user to modify the orientation (i.e., longitude and latitude) of the cones as well as their field of projection (i.e., volume and spread). You can find this example in Inviso_Cones.zip, which can be found at vraudio.org under Chapter 6.

On the other hand, if the listener node is outside the inner cone's field of projection, the user hears an indirect sound coming from the outer cone (like hearing someone talk in another direction). As a result, when the listener node moves outside of the inner cone's field of projection, the sound is not immediately cut off but rather faded out.

In Figure 6.8, we see the same listener node as Figure 6.6 moving on a trajectory. However, the two sound objects are now turned into unidirectional sources with cones. Whereas the cone on the left has a wider spread, the one on the right has a more focused directional pattern. The figure depicts a moment wherein the listener node is passing through the field of projection of the left cone, which will be clearly audible to the user. On the other hand, the right cone won't be audible in its full extent until the listener node passes through its field of projection. You can test the scene shown in Figure 6.8 by importing Inviso_Unidirectional.zip into Inviso. Try changing the longitude, latitude, volume, and spread properties of a cone to see how these affect its orientation, directivity, and audibility.

When using omnidirectional and unidirectional sounds in a game, it is essential to position the sound source at the visual mark where it is emitted from. For instance, when you are assigning footstep sounds to a character, place the sound source at the foot of the character. In traditional gaming, where the player controls a first- or third-person character on a 2D display, the sounds originating from a character can be placed at the root of the model. However, in VR, where the user has not only a first-person view, but also the ability to tilt their head or move it up and down freely, this kind of placement will create awkward spatial mismatches between a sound and the visual model it originates from.

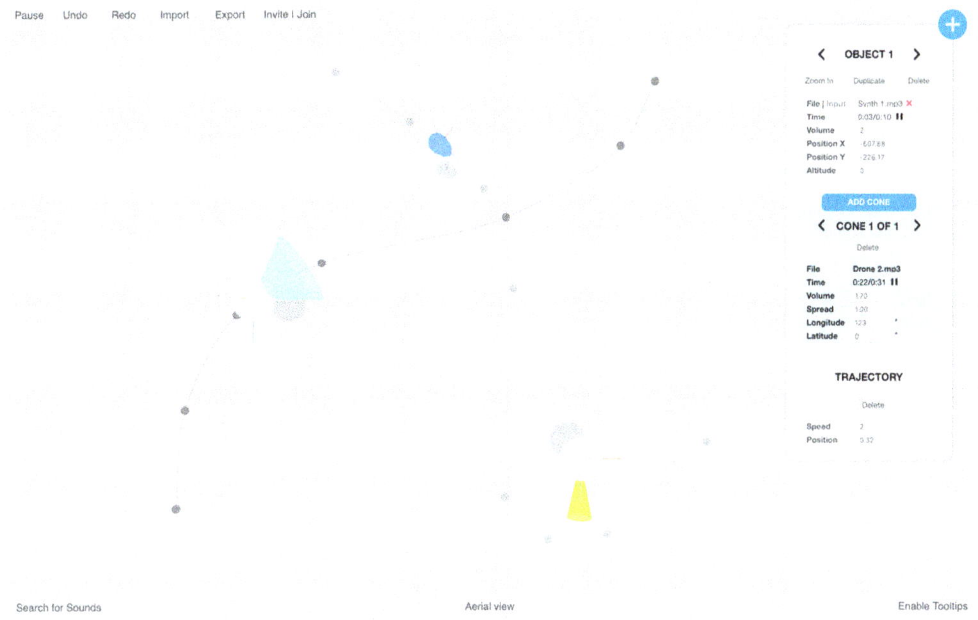

Figure 6.8 An Inviso scene similar to that in Figure 6.6. However, the omnidirectional sources have been transformed into unidirectional sources with a cone each. You can test this scene in Inviso by importing Inviso_Unidirectional.zip, which can be found at vraudio.org under Chapter 6.

Finally, **nondirectional sounds** are those that do not display any distinct spatial properties and are heard consistently regardless of the listener node's position in the scene. Nondirectional sounds offer a computationally inexpensive way of establishing ambience in a game. This could be the electrical hum in a factory scene or the sound of rain in an open-air environment. In traditional games, these sounds are delivered to the listener without dynamic spatial cues, serving as an acoustic background for spatialized sound sources. Nondirectional sounds are often triggered when the player enters a predetermined area, for instance, when they move from indoors to outdoors.

As we discussed in Section 5.4.1, head-locking a sound source will cause the sound to be anchored to the user's head and maintain its position relative to the user rather than virtual space (as illustrated in Figure 5.2). While it may sound intuitive to head-lock nondirectional sounds, this can be immersion-breaking since, even if we don't pay attention to the localization of ambient sounds in real life, we don't expect them to move with us as we move our heads. A work-around to this would be to render nondirectional sounds as Ambisonic files. This way, the user could still rotate their head without the nondirectional sounds moving with it. Although this will be computationally more expensive than monophonic playback, it will still be much more economic than composing an ambience out of individual spatial audio sources dispersed throughout the virtual environment. We will implement an example of this in our first case study in Section 6.7. Nevertheless, certain sounds that make up an ambience could still benefit from individual spatialization, especially if it corresponds to a dynamic background object. Furthermore, procedural audio can be used to animate nondirectional ambience. For instance, when the user moves through space, the ambience can be gradually filtered or modulated to give it some dynamism.

Another sound that is presented nondirectionally in traditional games is background music. Background music is often heard when the player enters a particular area or level in the game. It's presented without any spatial cues so that its audibility is not affected by the player's position as long as they remain in the area associated with the music. This clearly sets the music apart from the diegetic sounds in the scene. However, a head-locked presentation of background music can similarly be immersion-breaking in VR. A possible solution to this is to similarly mix the background music in Ambisonics. Nevertheless, we should check that the music is not perceived as originating from the scene as a diegetic source when mixed this way. This, too, can be immersion breaking as the music can come across as a phantom source in the scene. Furthermore, the space effects applied to the music track can conflict with the virtual space perceived visually.

Nondirectional sounds are useful for other sources that are meant to be non-diegetic (i.e., not part of the implied universe of a game environment). For instance, non-diegetic audio cues, like voice-overs and notification sounds are implemented as nondirectional sounds, suggesting that these sounds are part of an extradiegetic layer rather than a part of the game environment. However, it is also possible to change the diegetic disposition of a sound source by changing its spatial properties and move it from the game environment to an extradiegetic layer. As we have discussed in Section 5.4.1, these types of spatial manipulations can offer interesting narrative possibilities in VR. For instance, when telecommunicating with another player or an NPC in VR, their voice could be presented as a nondirectional source. However, as we move closer to them, their voice can gradually become a directional source localized in space as a navigational cue.

In Figure 6.9, the directional sound objects from the previous figures are replaced with nondirectional zones. In Inviso, these zones can be defined with free-hand drawings, which can then be adjusted using the control points of the spline curve that demarcates a zone. When the listener steps into one of these zones, the sound associated with it becomes audible monophonically without any localization cues. When the listener node passes through an overlapping area between two zones as seen in Figure 6.9, the sounds of both zones are heard. You can test the scene shown in this figure by importing Inviso_Nondirectional.zip file into Inviso. Try deleting the trajectory on the listener node and moving it with either your mouse or the W, A, S, D keys on your keyboard. Move in and out of zones and adjust zone shapes and parameters to see how these affect their audability.

Once you are done testing the Inviso scenes presented in this section, revisit the Inviso_Introduction.zip shown in Figure 6.5 to test a scene that incorporates all the directivity types that we covered here. Explore this scene by moving the listener node and observe how sounds with different directivity characteristics coalesce into a rich virtual soundscape.

6.5.3 Distance Modeling

Besides the position and directivity of a sound source, another factor that impacts its localization is how this sound propagates over distance. In Section 3.3, we discussed some fundamental properties of sound as it travels through the air, including the inverse square law, which states that the intensity of a sound is inversely proportional to the square of the

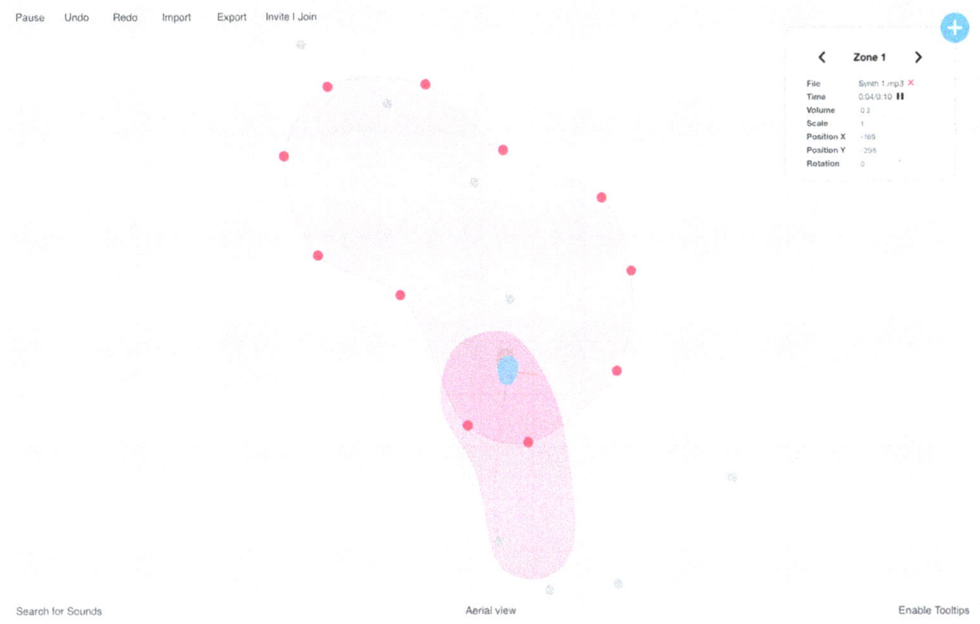

Figure 6.9 An Inviso scene where the listener node is crossing over two zones on a predefined motion trajectory. You can test this scene in Inviso by importing Inviso_Nondirectional.zip, which can be found at vraudio.org under Chapter 6.

distance it travels. This results in a steep roll off in sound intensity as the sound moves further away, with the roll off flattening out as the intensity approaches zero.

When we simulate sound propagation in VR, we can utilize an inverse square curve to model the behavior of real-life acoustic phenomena. However, we can also employ various attenuation functions besides this curve to adjust how audible a sound is at different distances. Some of the attenuation functions that we commonly encounter in game engines are logarithmic, linear, and custom curves, as illustrated in Figure 6.10. By leveraging these functions, we can create complex distance models for each sound source. For instance, in a scene with multiple NPCs and objects that emit sounds, a logarithmic function can be applied to sound sources that only require the player's attention when they are nearby. This way, the chatter of an NPC becomes audible only when it's nearby, rather than distracting the player by constantly occupying their listening space.

A linear function, on the other hand, could be applied to sound sources that are intended to draw the user's attention to distant events or set an overall ambience for the scene. Although approaching this source would increase its volume, it would be audible regardless of how far it is to the player. For instance, explosion sounds can be set to have a linear roll off, so that even when the player isn't near the explosion site, the sound that emanates from it is audible. If the player happens to be within or close to the explosion site, the sound becomes more prominent.

We can also define a custom function that allows an arbitrary relationship to be defined between the intensity of a sound source and the player's distance to it. An example of a custom attenuation function can be seen in Figure 6.10 (c), where a roll off at close

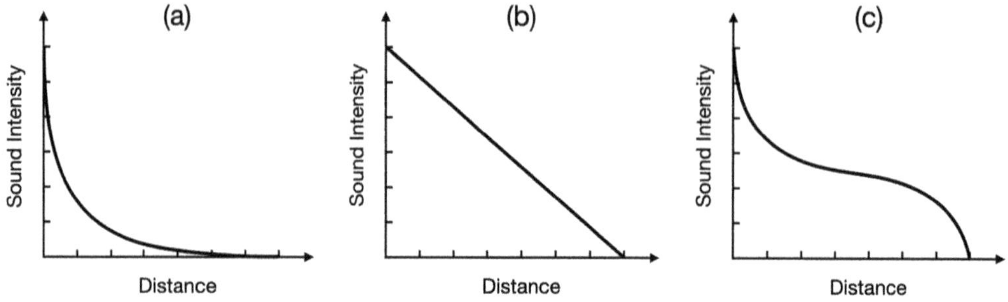

Figure 6.10 Attenuation functions where the vertical axis represents sound intensity and horizontal axis represents distance from the sound source. (a) Logarithmic function, (b) Linear function, (c) Custom attenuation function defined by the designer.

distances is followed by a flat region where the intensity of the sound doesn't change much with distance. This is then followed by a steep roll off at greater distances. This type of curve could be useful for making a sound consistently present within a particular range while making it more pronounced in close proximity. Modern game engines offer the ability to define attenuation curves with an arbitrary number of control points, allowing the creation of custom distance models.

In Section 5.4.4, we talked about heightening reality through sound for narrative purposes in VR film. The distance modeling in games can be used to this end by defining unrealistic attenuation characteristics that stray from a faithful simulation of acoustic phenomena. This may entail making objects hard to hear at close distances to draw the user closer in or making them highly audible even at far distances to amplify their role in the story. These types of deviations from reality can help engage the player or move the story forward. Moreover, humans already have a tendency to underestimate the proximity of a sound source at increasing distances (Fluitt, Letowski, and Mermagen 2014). In other words, we can perceive distant sources closer than they really are. Therefore, straying from a physically accurate distance model can in some cases more closely match our expectations.

6.5.4 Motion Parallax

Motion parallax is an effect wherein an object moving at a constant speed appears to move faster when it's closer to us than when it's farther away. For instance, when we are riding in a car on a highway, a car that zooms past us will appear to be moving much faster than a plane flying high in the sky despite the plane moving substantially faster than the car in relative speed. An analogous phenomenon occurs in the auditory domain (Genzel et al. 2018). Imagine a sound object moving around us in a circle. If the object moves closer to us while still maintaining a circular path (i.e., the radius of the circle is reduced), we perceive it as moving faster, even when its speed remains constant. The Inviso scene in Figure 6.11 offers a demonstration of this. Import Inviso_MotionParallax.zip into Inviso and compare the two objects by pausing one and playing the other in the control panel on the right.

We can leverage motion parallax by making objects move around the player at closer distances; the perceived increase in the speed of these objects due to their proximity can instill a greater sense of dynamism in the scene. By that same token, spatial audio

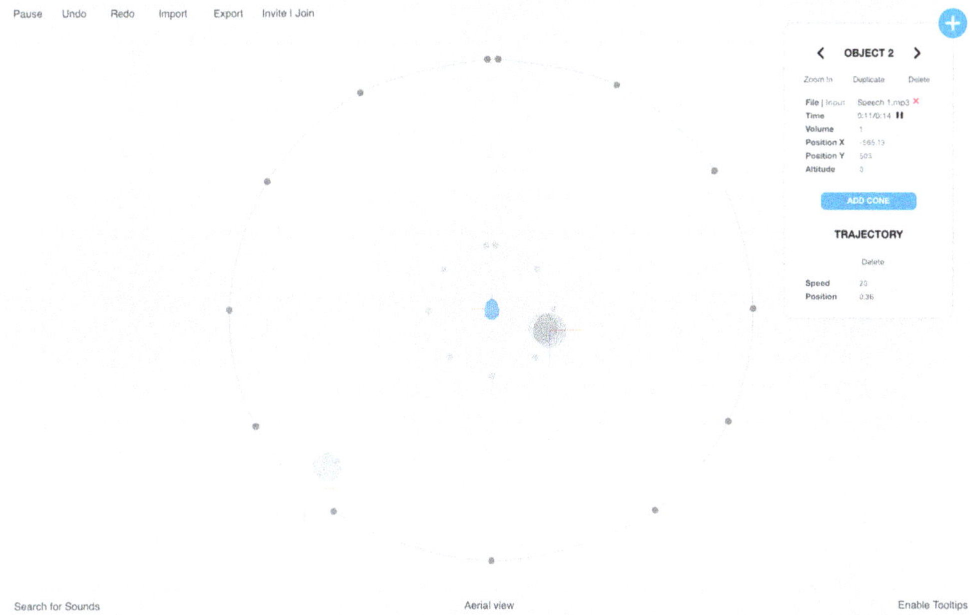

Figure 6.11 An auditory motion parallax demonstration in Inviso. In this image, the closer object is paused so that the user can hear the farther object alone. You can test this scene in Inviso by importing Inviso_MotionParallax.zip. Alternate between the two to see how the distance affects your perception of an object's speed.

elements that are too far away from the user may fail to evoke a sense of movement even though we would be incurring the same computational cost to spatialize them. These can be either sped up to make up for the motion parallax, or they could be made stationary or even converted to non-spatialized sounds depending on their function. On the other hand, we should also be cautious of this phenomenon as too many objects that are rapidly spatialized around the user can be disorienting. Furthermore, objects that move too rapidly and too close to the user may cause discomfort or induce motion sickness.

6.5.5 Room Effects and Occlusion

So far, we have been discussing the spatialization of sounds based on their positioning and propagation in free-field spaces. However, the perception of sound in space relies on not just its propagation but also the way it interacts with its surrounding environment and the objects in its path. Imagine a scenario where you are attending a piano recital in a concert hall. As the sound of the piano travels directly from the instrument to you, it will also bounce off all the surfaces surrounding it and reach your ear in the form of reflections that trail the direct sound. Not only that, but the reflections themselves will bounce off these surfaces and travel towards you. All these reflections will inform you about the size of the concert hall and the materials of its walls, which will govern the frequency response of the room (i.e., how it reflects and absorbs certain frequencies). These reflections determine the reverberant qualities of an acoustic space, which are critical to how we perceive and localize the sounds in it.

The reverberant qualities of a virtual space can be simulated in several ways. A traditional reverb effect based on delay networks can be used to match the acoustics of the space based on its visual size and properties. We can adjust parameters like pre-delay amount and decay length to manipulate the perceived size of the acoustic space. We can then use input and output filters to modify the absorbent qualities of this space. Another artificial reverb technique that we discussed in Section 3.5 is convolution reverb, which relies on the impulse response of real-life environments. We could use the impulse response of a real space that matches the acoustic properties of the target virtual environment for a convincing reverb effect.

Moreover, the acoustics of a virtual space can be modeled based on its properties. This type of room modeling can be achieved by either solving acoustic wave equations (Gazdag 1981) or by simulating geometrical acoustics by ray tracing (Savioja and Svensson 2015). While the former can yield more accurate results, it is also more computationally intensive. On the other hand, although ray propagation doesn't fully capture the behavior of sound waves, geometrical acoustics can be an efficient way of creating realistic room effects. This is why modern game audio implementations increasingly employ ray tracing to simulate room acoustics. For instance, the Steam Audio spatial audio library,[2] which we will utilize in our first case study in Section 6.7, offers both precomputed (i.e., baked) ray tracing and real-time rays traced from the listener node to calculate reflections and apply reverb to the sound sources in a game. Another spatial audio library called Resonance Audio (Gorzel et al. 2019) similarly utilizes ray tracing to precompute the reverberation in a game environment based on probes placed in the scene. Baking the reverb eliminates the need to perform real-time ray tracing during run time. This approach offers the realism of geometrical acoustics while saving on processing power.

Ray tracing has other applications in game audio besides reverb modeling. Imagine another scenario, where your roommate plays music through loudspeakers in their room. The wall between your rooms would impede this sound's propagation towards you. By the time it reaches your ears, the sound would not only be quieter but also somewhat muffled because the wall would absorb high frequencies more effectively than it does low frequencies. We refer to this phenomenon as acoustic occlusion. To implement this in game audio, a ray is cast between the listener node and the sound object. If a game object stands in the way of this ray, it is factored in as an occlusion. The precision with which this occlusion is computed will depend on how much computational resource can be allocated to it. At its simplest, the occlusion will cause a low-pass filter to be applied to the sound source. However, more elaborate occlusion behaviors can incorporate the geometry and material of the occluding object to determine how exactly a sound would be absorbed or transmitted by this object. Reflections around this occluding object can also be calculated for a more realistic acoustic modeling of the virtual space.

In the following sections, we will explore practical applications of the spatial audio topics that we covered here. We will look at how concepts like source directivity, distance modeling, and occlusion are implemented in game engines. We will then apply these to the design of two VR music games. To this end, we will use two of the leading game engines used in game and VR development, looking at the individual workflows they offer for spatial audio and, more broadly, VR design.

6.6 Unity Overview and Audio Features

A game engine can be viewed as an integrated development environment (IDE) that incorporates various functions like scene design, lighting, physics, animation, scripting, file management, and debugging. These features make game engines powerful tools for building multimedia environments, including those used in VR experiences. One of the leading game engines today is Unity, which is widely used to build games for various operating systems and gaming consoles. It is also employed in the development of web, mobile, and VR applications.

Unity offers a C# scripting API (application programming interface) that enables designers to implement game mechanics and interactions.[3] Scripting is essential for developing dynamic behaviors and rule systems that govern how game objects operate and how the user interacts with them. Before we get into our first case study, which will utilize C# scripting among other functionality of the engine, we will first take a look at some of the fundamental features of the Unity Editor. We will then shift our focus to Unity's audio implementation.

The default layout of the Unity game engine consists of four primary views as shown in Figure 6.12. At the center is the **Scene View**, which gives a visual representation of the 2D or 3D game environment and the various elements that populate it. These elements are referred to as game objects. When we press the play button at the top of the screen, the Scene View switches to the **Game View**, which shows a playable runtime version of the scene.

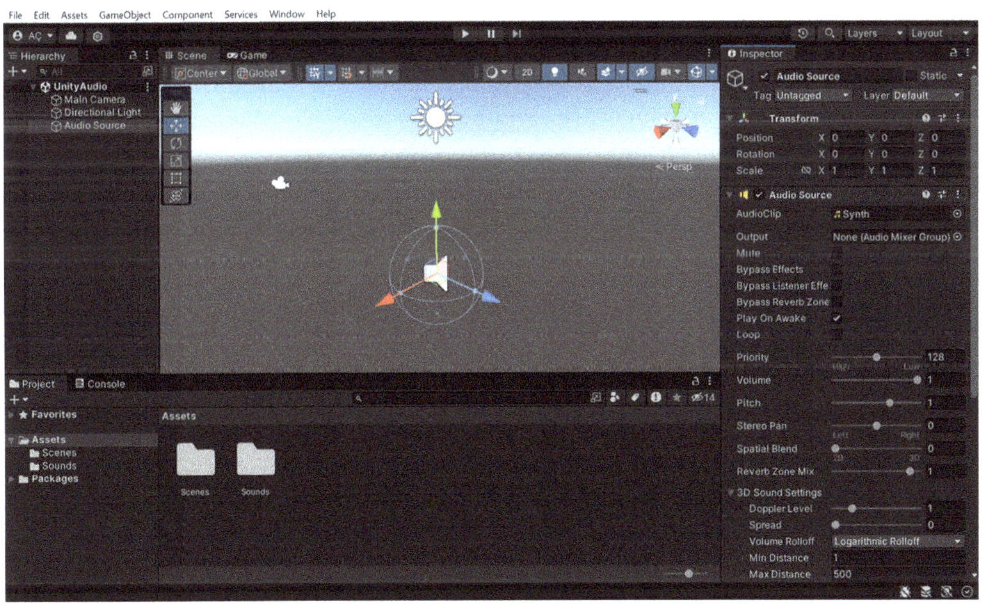

Figure 6.12 A Unity project with a single Audio Source added to the scene. The Unity editor layout includes the Scene View at the center, the Hierarchy Window on the left, the Project Window at the bottom, and the Inspector Window on the right.

The panel to the left of the Scene View is the **Hierarchy Window**, which displays a list of game objects that are currently in the scene. This window is named as such because it gives a hierarchical view of the game objects, which can be nested into each other. When an object is placed inside another object, it becomes a "child" of the "parent" object. This way, a scene can be given a logical structure that reflects the relationship between the game objects. If, for instance, a scene involves a room object, all the furniture objects can be nested in it. This not only makes the Hierarchy Window easier to navigate but also allows a parent object to be moved in the scene along with all its nested child objects without disturbing the spatial relationship between them.

The **Project Window** at the bottom shows the assets (e.g., models, materials, textures, scripts, audio files) that are associated with the current project. The folder structure shown here mirrors the Unity project folder in the computer's file system. Any changes made in the Project Window are reflected at the system level as well (i.e., deleting an asset from the Unity Project Window will also delete it from the computer). An asset can be dragged from the Project Window into the Hierarchy Window or the Scene View to create a new game object. Alternatively, an asset can be attached to an existing game object, for instance, when a texture is applied to a model in the scene.

The **Inspector Window** to the right of the Scene View shows the components of the currently selected game object or asset. A component can pertain to the transform (i.e., position, rotation, and scale) of an object, the model that gives it a form, its material settings, the scripts that imbue it with custom functionality, or the sounds attached to it, among other things. In combination, these components define the properties and behaviors of a game object.

6.6.1 *Audio Components in Unity*

The audio components in Unity can be grouped under three categories, namely the listener, source, and filter components. The **Audio Listener** component is Unity's implementation of a listener node and serves as the player's ears into the virtual space. In first-person games, the Audio Listener component is attached to the camera, whereas in third-person games, it is attached to the player character. Therefore, as your character moves through the virtual environment, you hear the sounds in it from the character's perspective. In VR, the Audio Listener component is attached to the camera object that generates the visual feed for the head-mounted display. As the user moves within virtual space with 6DoF, the camera object and its Audio Listener component follow these movements. Note that, there should only be a single Audio Listener component active in the scene at any given time. If there are multiple Audio Listener components in a scene, the debugger will throw a warning indicating that these should be reduced to one, which Unity will take as a reference point when processing the audio spatialization.

The primary audio component in Unity is called the **Audio Source**. This component can be attached to any existing game object from the object's Inspector Window. However, empty game objects that consist of an audio component without any visual counterparts can also be created. This can be achieved in three ways: 1) by pressing the plus button at the top left corner of the Hierarchy Window and scrolling down to the Audio tab in the pop-up menu; 2) by right-clicking anywhere in the Hierarchy Window and selecting the Audio tab in the pop-up menu; 3) by accessing the GameObject menu in the task bar at the top of the editor and selecting the Audio tab. These actions are equivalent to creating an empty game object and attaching an Audio Source component to it.

An object with an Audio Source component will serve as a point source (i.e., an omnidirectional sound). In other words, the sound of this source will propagate evenly in all directions in a spherical pattern. At the time of this writing, Unity does not offer a native implementation of sound cones. However, we will soon discuss the use of third-party libraries that enable the use of directional sounds in Unity.

Figure 6.13 shows a close-up of the Audio Source component in a game object's Inspector Window. In this component, we can attach an audio file to the associated game object and control which Audio Mixer Group its sound will be routed to. If no Audio Mixer Group is defined, the sound of the object will be routed to Unity's main audio

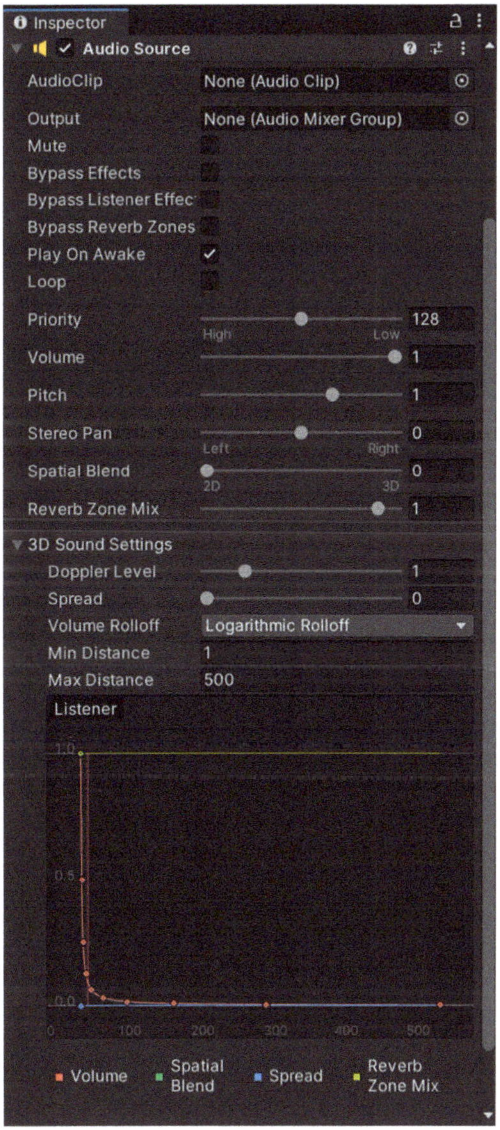

Figure 6.13 A close-up of the Audio Source component.

output, which is the same as your computer output. The following checkboxes allow us to control various runtime attributes of the Audio Source. Using these, we can mute a sound during gameplay, bypass audio effects applied to it, allow the sound to interact with the reverberant zones in the scene, control whether the sound should start playing when the game is launched, and determine whether it should play only once or repeatedly.

The priority parameter allows us to control how computational resources are allocated to audio playback. In earlier versions of Unity, the engine imposed a limit on the number of individual sound sources (voices) that could be played back simultaneously. Starting with Unity 5, this limit was turned into a customizable parameter constrained by the hardware capabilities of the target system. The priority control allows certain sounds to be favored over others when a resource limit is reached. This parameter works in a range between 0 to 256, where 0 indicates the highest priority and 256 indicates the lowest. When a hardware limit on the number of voices is reached, Unity uses priority values to decide which sounds should continue playing back and which ones should be stopped. When two sources are of the same priority, Unity relies on an "audibility" metric based on the source's distance from the listener node and its loudness level. In many cases, it is advisable to leave diegetic sound sources at the same priority level and allow the audibility mechanism to handle the voice allocations. That being said, non-diegetic sounds like music, essential dialogue, narrative voice-overs, and user interface alerts can be given a higher priority to ensure that these are not cut off under any circumstances.

The volume control allows a sound source to be attenuated, with 1 indicating no attenuation (i.e., unity gain) and 0 indicating full attenuation (i.e., muting). This and other parameters of the Audio Source can be automated with the curves shown at the bottom of the component under 3D Sound Settings. This way, the volume can have a dynamic behavior based on the distance between the sound source and the listener node. All the parameters of an Audio Source can also be controlled during gameplay using scripts or animation curves.

The pitch parameter defines the speed of audio playback with 1 corresponding to the original playback speed of the source. Turning up this value will lead to an increase in the playback speed, and, therefore, a rise in the perceived pitch of the source. For instance, when a source is played back twice as fast, its pitch will be shifted up by an octave, whereas halving the original speed will shift the pitch down by an octave. With this octave relationship in mind, the twelfth root of two (~1.05946) can be used as a scalar to control the pitch of a sound source in semitones, achieving musical intervals on a standard twelve-tone musical scale.

Besides the volume and pitch parameters, the Audio Source component also offers control over the spatial properties of a sound source. The stereo pan parameter can be used to adjust the balance between the left and right output channels. The spatial blend parameter controls how much a source is influenced by Unity's built-in 3D audio engine. Lastly, the Reverb Zone Mix parameter controls how much of a sound is fed into the reverb effect when the game object coincides with Audio Reverb Zones—another audio component type which we will discuss shortly.

For sources that have the Spatial Blend parameter turned up, a collection of 3D Sound Settings offer additional control over sound spatialization. The Doppler Level controls how much a moving source is affected by a Doppler effect simulation. The Doppler effect is an acoustic phenomenon that we encounter in real life when we observe a moving sound source. When a source is moving towards us, its pitch is perceived to be increasing whereas the opposite occurs when it's moving away from us. This is because, as the source approaches us, each cycle of its sound wave is emitted from a position closer to us than the previous cycle, causing the perceived frequency of the wave to be higher than the emitted. The same principle works in reverse when the sound moves away from us. A common example of this is illustrated in Figure 6.14. When an ambulance is passing by us, the perceived pitch of its siren goes up until the vehicle is closest to us; the pitch then starts to decline as the ambulance moves away from us. While a subtle amount of Doppler effect can increase the perceived realism of a moving virtual object, overdoing this effect can cause exaggerated pitch shifts that adversely affect the realism of the sound source.

The Spread parameter controls how wide a sound source appears, even before any reverberation or room effects are applied. By adjusting this parameter, the size of a dry sound source can be manipulated by allowing it to bleed into the opposite channel in a stereo setting. In multichannel setups, increasing Spread causes the source to bleed into other channels, making it occupy a wider space in the audio field.

The Volume Rolloff parameter controls the type of curve applied to the distance model of a sound source. This curve determines how the loudness of a sound should change in relation to its distance from the listener node. To reiterate our earlier discussion on distance modeling, the Logarithmic Rolloff option approximates the inverse square law, where the intensity of a sound source is inversely proportional to the square of the sound's distance from the listener. This option is best suited for sound sources that need to be audible when the player character is near the game object that emits this sound. Linear Rolloff, on the other hand, is useful when a sound source should decrease in volume as the user moves away from it but should still be audible even at a distance. Finally, the Custom Rolloff function allows us to create our own distance curve. The graph at the bottom of the 3D Sound Settings section allows for the manipulation of these three types

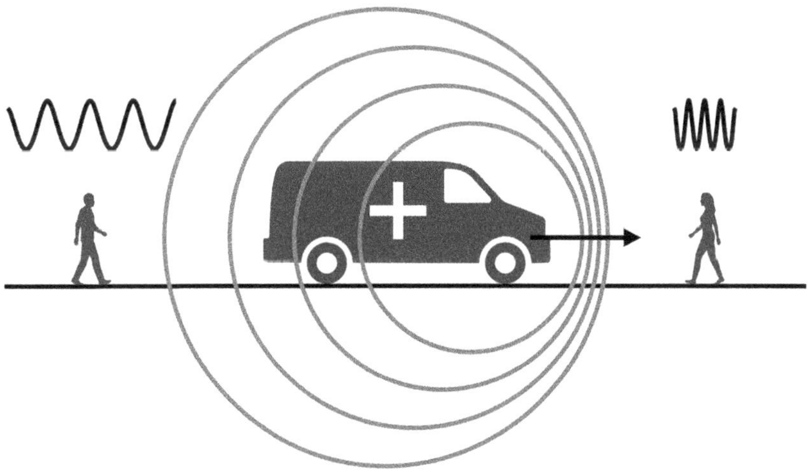

Figure 6.14 An illustration of the Doppler Effect based on the ambulance example. As a sound source moves towards a listener, its perceived pitch gradually goes up. Conversely, as it moves away from the listener, the perceived pitch goes down. These changes in pitch contribute to our perception of sound in space by indicating the speed and direction of a moving sound source.

of volume curves, as well as the automation of the Spatial Blend, Spread, and Reverb Zone Mix parameters. These parameters are represented along the horizontal axis, while the distance between the sound source and the listener node is plotted on the vertical axis.

Another important audio component in Unity is the **Audio Reverb Zone**, which allows us to designate areas wherein a reverb effect is applied to Audio Sources. An Audio Reverb Zone contains two regions: an outer region where the reverb is gradually introduced to the Audio Source as the player enters the zone, and an inner region where the reverb effect is applied to the source in full force. The Audio Reverb Zone comes with several presets that define different spaces such as cave, hangar, and auditorium. We can also choose the "User" preset and change the reverb parameters manually. These parameters include decay time, density, and diffusion as well as level controls for early reflections, room effects, and late reverberation.

Finally, the third group of audio components in Unity is **Audio Filters**. These include some common audio effects such as chorus, distortion, echo, reverb, high-pass filter, and low-pass filter. While these can be used for applying a constant effect to a sound source, these components become more effective when they are dynamically controlled via scripting to procedurally manipulate sounds.

6.6.2 Extending Unity's Spatial Audio Features

Although Unity's built-in audio features enable the implementation of basic spatial audio scenes, it lacks some of the more advanced features that can be expected from modern spatial audio tools, such as source directivity, occlusion, and binaural rendering. In this section, we will explore the use of a third-party library to extend the built-in spatial audio capabilities of Unity.

Most VR hardware manufacturers offer software applications or libraries to facilitate the integration of their products with game engines. These libraries can also include spatial audio plugins that extend the native audio functionality of the game engine. However, these are often purpose-built for a particular hardware and are not intended to interface with devices from other manufacturers. There are also third-party spatial audio libraries, such as Steam Audio and Resonance Audio, that can be used across numerous hardware and software platforms, including Meta Quest, iOS, Android, and web browsers. Here, we will focus on Steam Audio, which offers a robust set of spatial audio features. The Steam Audio plugin is developed by Valve, which is a game design and distribution company. As makers of the Index VR system and the highly popular VR game *Half-Life: Alyx*, Valve is heavily invested in VR as a gaming platform. In 2024, the Steam Audio library was made open source to allow developers to change or build upon its source code.

To use the Steam Audio library, its Unity package needs to be imported into the current project from the menu bar under "Assets > Import Package > Custom Package...". Once the library is imported, the Spatializer Plugin setting under "Edit > Project Settings... > Audio" needs to be switched to "Steam Audio Spatializer." If the project utilizes Ambisonic content, the Ambisonic Decoder Plugin setting should also be changed. The Steam Audio package comes with several components that can be attached to game objects to give them spatial audio features. Among these are Steam Audio Source, which expands the built-in Audio Source component; Steam Audio Listener, which allows a game object to be assigned as a Listener Node for Steam Audio Sources; and Steam Audio Geometry, which tags a model as an acoustic geometry that contributes to occlusion and reflections. Although there are other scripts and functionality offered by the Steam Audio Library, let's focus on these three for now.

First off, the Steam Audio Listener allows us to adjust certain properties of the acoustic simulation, such as the application and baking of reverb in the scene. This script should be attached to the camera object that provides the visual feed to the VR headset. The same object will also need to have the Unity Audio Listener component attached to it.

Figure 6.15 shows a Unity project, where the Steam Audio Source is attached to one of the game objects. Note that the Steam Audio Source relies on an existing Audio Source component. When Steam Audio (or another external library) is selected as the Spatializer Plug-in in the Project Settings, this Audio Source component will show an additional parameter named "Spatialize." This box must be checked for the Steam Audio spatialization to be applied to the Audio Source. Additionally, the "Spatial Blend" parameter needs to be turned all the way up for the spatialization to take full effect.

On the right-hand side of Figure 6.15, we see the Steam Audio Source component with most of its features enabled. The Direct Binaural option under HRTF settings switches the output from panning-based spatialization to binaural rendering. The interpolation can be set between Nearest and Bilinear, with the latter offering a more detailed HRTF interpolation with an increased computational cost.

Once Distance Attenuation is enabled, the user can pick between a Curve Driven option, which adheres to the Volume Rolloff curve defined in the Audio Component, or a Physics-based option, which relies on the Inverse Square Law. Here, the user can also enable Air Absorption. When the Simulation Defined option is selected, air absorption

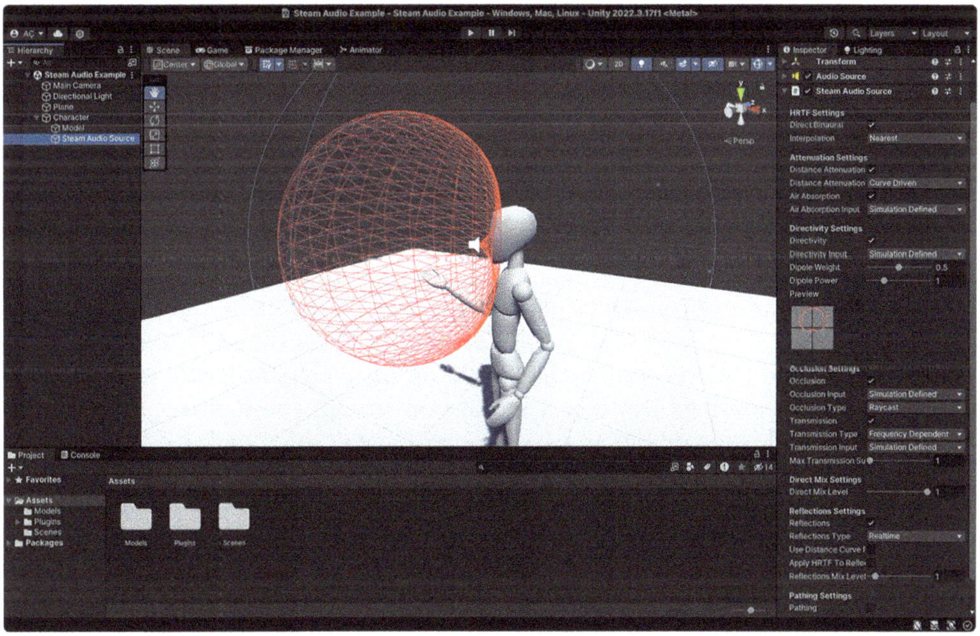

Figure 6.15 View of a game object with a Steam Audio Source component. When this object is selected, the polar pattern of the sound source is displayed as a 3D mesh in the scene.

relies on a physics-based model with an exponential falloff, where the high frequencies are attenuated with distance more rapidly than the lows. The User Defined air absorption allows the low, mid, and high frequencies to be attenuated individually with a script.

Directivity is one of the crucial features that Steam Audio brings to Unity. When the Simulation Defined setting is used, the Dipole Weight value allows the directivity to be changed from an omnidirectional (monopole) pattern at 0 to a figure-8 (dipole) pattern at 1. When set to 0.5, the directivity pattern becomes a cardioid as seen in Figure 6.15. At this setting, the audio source becomes unidirectional. The Dipole Power value controls the sharpness of directivity. When this is set to 0, directivity takes an omnidirectional shape regardless of the Dipole Weight, whereas 1 gives the narrowest directivity available. The User Defined input setting allows a custom directivity pattern to be implemented with a script. When Simulation Defined is selected, the pattern is visualized with a red mesh as seen in Figure 6.15. In this figure, the mesh indicates the directivity pattern of a dialogue sound. Notice how the Steam Audio Source component is aligned with the mouth of the humanoid model, following our previous recommendation that sound sources should be positioned where they are meant to originate from in the environment.

The Occlusion Settings is another important feature of the Steam Audio Library, which allows models in the scene to block or transmit sounds. This is where the Steam Audio Geometry script comes into play. In the current workflow, we attach this script to every object in the scene that is meant to cause acoustic occlusion. We then export the active scene from the Steam Audio Menu in the task bar. This creates an occlusion settings file, as well as a "Steam Audio Static Mesh" game object that references this file. Every time a new object is identified as a Steam Audio Geometry with this script, the settings file needs to be re-exported for this object to contribute to acoustic occlusion.

Going back to the Occlusion Settings on the Steam Audio Source component, when the Occlusion Input is set to Simulation Defined, the default Occlusion Type is Raycast, where a single ray is cast from the Steam Audio Listener to the Steam Audio Source. If a Steam Audio Geometry happens to be in the way of this ray, occlusion is applied to the sound. When the Occlusion Type is set to Volumetric, multiple rays are cast between the listener and the source to apply the occlusion effect more gradually, albeit at a higher computational cost.

With Transmission disabled, occlusion will fully block a sound, which may create an unnatural acoustic response. Revisiting our discussion on acoustic occlusion in Section 6.5.5, remember the scenario where your roommate plays music in their room. Unless your rooms are acoustically isolated from each other, there would be some degree of sound transmission between your rooms based on the material, thickness, and insulation of the wall separating them. The Transmission settings allows this phenomenon to be simulated by allowing objects to pass through some of the sound that they occlude. The Transmission Type can be frequency independent, where an attenuation is applied to the entire spectrum, or frequency dependent, where the low, mid, and high frequencies can be attenuated separately. Much like the Air Absorption and Directivity Input settings, the Transmission Input can be either Simulation Defined (i.e., physics-based) or User Defined with the low, mid, and high frequency transmissions controlled with a custom script.

Finally, the Reflection Settings enable room effects to be applied to a sound source. Although Unity's Audio Reverb Zones can also be used to this end, Steam Audio calculates reflections that bounce off of surfaces to simulate a room effect based on ray-tracing. While this will be computationally more intensive than a more traditional reverb effect, it will be more closely associated with the virtual space. The library offers various material types, such as concrete, glass, wood, and rock, that can be applied to Steam Audio Geometries. Using these, the reflective properties of different surface materials can be incorporated into the simulation of room effects in real time.

6.7 Case Study 1: *The Fountain*

In our first case study, we will discuss the design and implementation of a VR music game called *The Fountain*, which is created in Unity using Steam Audio and a range of sound sources from Ambisonic recordings to procedural audio. Inspired by modular synthesizers, *The Fountain* exploits the sense of scale in VR as the player interacts with various synthesizer components, such as oscillators, filters, and effects. More specifically, it reimagines common electronic music interactions, like moving a fader or rotating a knob, to examine the embodied relationship between a performer and their instrument. In doing so, *The Fountain* aims to expose the inner workings of a synthesizer and allow the player to explore its components individually. As seen in Figure 6.16, the player is presented with a virtual park, furnished with the interface elements of a synthesizer. The scale of these ranges from that of their real-life counterparts to the size of a building, dictating the mode and precision of interaction. The project files for this case study can be accessed at vraudio.org under Chapter 6. The reader is invited to download the project and explore its implementation in greater detail as they read through this section. Both this and the next case study is based on material that we previously reported at the Audio Engineering Society International Conference on Audio for Games (Çamcı 2024).

Figure 6.16 A screenshot of *The Fountain*. The image shows a view of the park from afar. At the center of the virtual space is a "fountain," where the player can control the overall volume of the synthesizer's output.

6.7.1 *Game Mechanics*

The player interacts with the synthesizer using virtual buttons, knobs, sliders, and switches that control the sound engine in a manner that is similar to using a MIDI keyboard or controller. However, these interface elements are dispersed throughout the virtual environment. Moreover, they are scaled in ways that dictate how they can be interacted with. For instance, the piano-style keyboard that the player is presented with at the beginning of the game is scaled similarly to a physical keyboard. Therefore, the user interacts with it using their fingers. On the other hand, the knob that controls the filter cut-off is 8 feet in size, requiring the player to use their arms to grab onto the knob and pull to rotate it.

Navigating the virtual space is another core mechanic of the game based on the use of spatial audio. Alongside the main output of the synthesizer, the player can hear the oscillator, filter, envelope, and delay components of the synthesizer in isolation. The directivity patterns and distance models applied to these sounds allow the sound of the park to evolve as the player moves through it. Combined with Ambisonic recordings and ambient drones, these sounds prompt the player to observe the virtual space from different vantage points.

6.7.2 *Audio Implementation*

The audio implementation of *The Fountain* relies on a mixture of fixed and procedural sounds with dedicated spatialization characteristics. Whereas some of these sounds play continuously to serve an ambient function, others are triggered or synthesized based on user input.

6.7.2.1 *Procedural Audio*

For procedural sounds, we utilized Pure Data (Pd), which is an open-source audio programming environment.[4] Pd has been previously utilized in video games for procedural audio.[5] Although Pd is a stand-alone application, it has an embeddable version called libPd, which can run on various hardware and software platforms. To run Pd natively in a game engine, we use the LibPd Unity Integration library,[6] which is an adaptation of libPd for the Unity API.

The LibPd Unity Integration library allows Pd patches to be run in Unity through a "LibPd Instance" component that can be attached to game objects. Adding this component to an object automatically creates an Audio Source component on the same object. The LibPd Instance component is followed by a Steam Audio component, which spatializes the output. Once a LibPd instance is created, other game objects can reference this instance to communicate with the Pd patch. This way, a game object can control the procedural audio patch attached to another other game object. The patches are opened and edited in the Pd environment. Double clicking a patch in a LibPd Instance component will launch this environment as a standalone application. However, the Pd environment is not necessary to run or build the game as the patch runs natively inside Unity.

The primary sound source in *The Fountain* is a square-wave subtractive synthesizer with an arpeggiator and an end-of-chain delay module that adds rhythmic and timbral

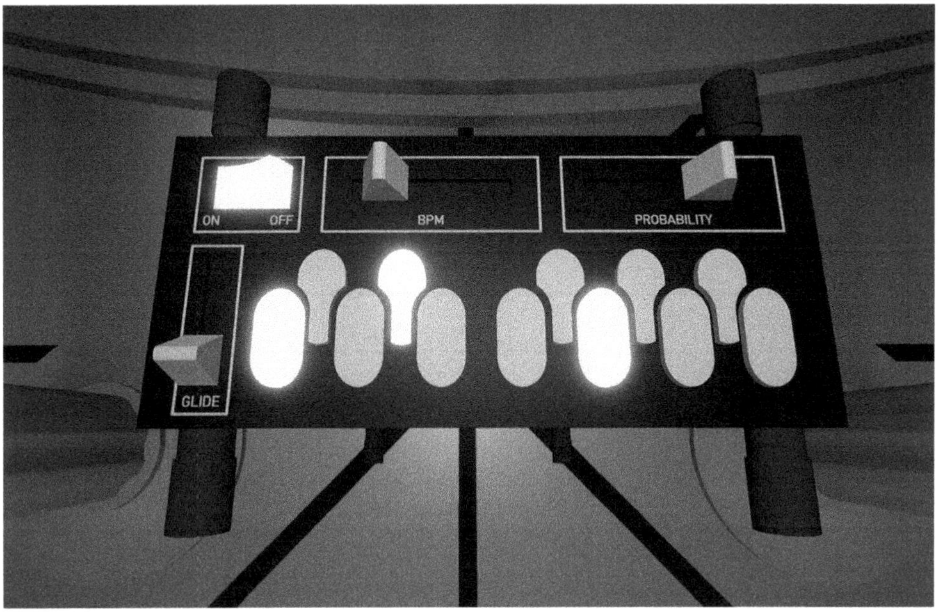

Figure 6.17 A close up image of the virtual arpeggiator placed atop a keyboard stand in *The Fountain*. The interface affords control over on/off state, note selection, the glide amount between notes, beats per minute (BPM), and the probability of a note to be played on each beat. The interface is scaled to the size of a real-life keyboard.

variety to the synthesizer's output. As shown in Figure 6.17, the arpeggiator, which is scaled to the size of a comparable physical interface, allows the player to control the on/off state, beats-per-minute (BPM), the probability of a note to be played on each beat, the notes to be included in the arpeggiation, and the glide amount between the notes.

The large knob placed atop the fountain functions as the main volume knob as well as the sound source for the synthesizer's final output. The interface elements for the remaining components of the synthesizer (i.e., oscillator, filter, envelope, and delay) are located in four quadrants around the virtual fountain. The user can hear and control a component by walking towards the quadrant dedicated to it. For instance, in the Oscillator quadrant, the player hears the raw output of the square-wave oscillator before any filtering or amplitude envelope is applied to it. In this quadrant, the player can control the frequency, pulse-width modulation (PWM) rate, frequency modulation (FM) rate, and FM depth parameters of the oscillator.

Figure 6.18 shows the architecture of the subtractive synthesizer implemented in Pd. Each box in this figure represents a component of the synthesizer, displaying the parameters that are exposed as interface elements in the game. The audio flow between the various components of the synthesizer is illustrated with solid arrows, whereas the control flow is shown with dashed arrows.

Instead of having different Pd patches for the individual components, the synthesizer relies on a single Pd patch with multiple outputs that monitor different parts of the synthesizer topology. For instance, when the patch is attached to the oscillator quadrant, the

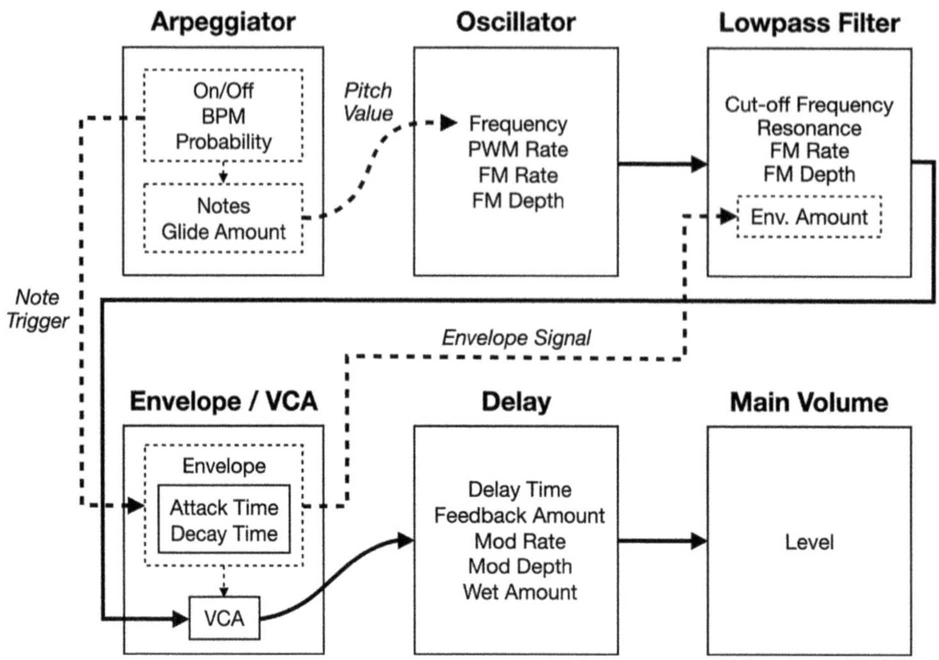

Figure 6.18 Architecture of the synthesizer designed in Pd, showing the individual components with their parameters that are exposed as interface elements in the game. The solid arrows indicate audio signal flow, whereas the dotted arrows indicate control signal flow.

output that monitors this portion of the synthesizer is activated. This ensures that any changes to the synthesizer can be applied to a single patch and propagate through all of its instances in the game.

6.7.2.2 *Fixed Sounds*

The virtual park in the game is populated with synthesizer modules surrounded by a topographic landscape. While the overall imagery evoked by these elements is abstract, the park metaphor, the cloudy sky, and the mountainous landscape off in the distance still instill a sense of being outdoors. This quality of the visual environment is supported by an Ambisonic recording of a real-life park, which was captured on location with a Sennheiser Ambeo and a Sound Devices MixPre-6 field recorder.

As we discussed in Section 6.1, recordings made with a single Ambisonic mic will offer 3DoF, whereas a game environment is 6DoF. In other words, while the player can move through the game environment, they can't move through the soundfield captured with the Ambisonic mic. To address this, in our recording of the park, we intentionally captured far-field sounds that instill a sense of orientation in space but are neither moving nor easily localizable. This allows us to attach the Ambisonic recording to the listener node, so it moves with the player. While the Cartesian coordinates of the soundfield follows the player position, its pitch, yaw, and roll remain fixed relative to the virtual space. To achieve this, we use a script to apply the position property of the player's transform to a dedicated Audio Component that loops the Ambisonic recording, while keeping the rotation property of this component unaffected by the rotations of the player. This way, the user can constantly hear the soundfield and rotate their head within it. It is worth noting that if the soundfield recording included any near-field sounds, mapping its position to that of the player could break immersion as these sounds would have suggested nearby objects that follow the player without corresponding visual cues.

For Unity to recognize this audio file as an Ambisonic recording, we select the file in the Project Window. This brings up the file's settings in the Inspector Window. Here, checking the "Ambisonic" box allows Unity to process it accordingly. For the binaural decoding of this file, we add the Steam Audio Ambisonic Source component to the game object which has the Ambisonic file as an Audio Source. We previously mentioned that we need to check the "Spatialize" button in the Audio Source component for the Steam Audio spatialization to take effect. However, with an Ambisonic recording, since the spatialization is already encoded into the B-format file, we must uncheck this button, as we don't need to apply another layer of spatialization to the source material.

For the abstract drone that represents the topographic mountains afar, we use evenly spaced Audio Components as emitters with an omnidirectional pattern and a custom logarithmic volume roll off. The distance model allows these sounds to have a wide reach but a substantial increase in volume when the player is close to them. As a result, the user hears a subtle ambient drone while close to the fountain. As the user moves away from the park towards the mountains, the drone sound intensifies and overpowers the sound of the synthesizer, indicating to the player that they are leaving the audible range of the synthesizer.

Finally, the buttons of the arpeggiator have a subtle click sound to confirm player interactions. The click is played from a single audio file that contains a one-shot sound recorded with a hardware synthesizer. However, each interaction causes a slight randomization in the pitch of this sound so that it doesn't feel overly repetitive. This also allows consecutive interactions to stand out as discrete events.

6.7.2.3 Spatialization

The sound spatialization in *The Fountain* is achieved with the Steam Audio library, which we showcased earlier in Section 6.6.2. The main output of the synthesizer is projected with an omnidirectional pattern from the fountain (i.e., the center of the scene). The distance curve applied to the attenuation of the sound gives it a relatively quick roll off. Therefore, the main output is only heard when the player is closer to the center of the scene within the area demarcated by the four quadrants. This configuration allows the main output to cross-fade into the sounds of the quadrants as the player moves away from the center. Each quadrant's sound is spatialized with a sound cone pointed towards the fountain. However, this cone has a relatively wide field of projection and a dipole weight that allows the sound to bleed in the opposite direction. As a result, while the sound is somewhat focused in the direction of the fountain, it's still audible from other parts of the scene. When this directivity pattern is paired with Steam Audio's occlusion features, traveling through the park results in interesting variations in the localization, attenuation, and filtering of the sounds from the four quadrants.

Scripting

While C# scripting is beyond the scope of this book, there are numerous resources that deal with this topic including online tutorials provided by Unity[7] and prior literature (e.g., Murray 2021). Nevertheless, *The Fountain* relies on several C# scripts that implement the interactions between the player and game objects. You can find these scripts in the project files linked earlier. You will notice that these scripts are heavily commented with descriptions that explain their functionality. We advise our readers to peruse these inline comments, even if they don't feel proficient enough to write their own scripts. For individual developers, C# scripting is already an essential skill for working with Unity. However, even when a VR audio designer operates within a team where they aren't tasked with scripting, it can be useful to be at least conversant with it. Understanding how scripting ties together the elements of a game will not only facilitate the audio designer's communication with the rest of the team but also inform their work in terms of the possibilities in audio implementation.

6.8 Unreal Overview and Audio Features

Another leading game engine today is Unreal, which was originally developed by Epic Games to create the game Unreal. Over time, it has evolved into a general-purpose engine integrated in other games. Today, Unreal Engine is used in a range of applications from video games and VR to car entertainment systems and virtual production. While it shares many features with Unity, it also offers advanced systems for virtualized geometries, lighting, character design, and sound. In recent years, Unreal Engine has gained popularity in the movie industry for virtual production, where rendered environments and visual effects are displayed on large-scale LED screens to serve as dynamic backdrops for film sets.[8]

In Unreal Engine, custom features like game logic, actor behaviors, and user interactions can be implemented using a dedicated C++ API. The engine also provides a visual scripting environment called Blueprints, which allows these game features to be built with node graphs. While the C++ API allows complex game systems to be implemented with efficiency, Blueprints are more accessible to novice users and those who do not have prior experience with C++ or other text-based programming languages. Nevertheless, Blueprints can also be used in the design of complex systems, albeit with some computational overhead.

Before we dive into Unreal Engine's audio functionality, let's explore some of its higher-level features. Looking at the engine's user interface in Figure 6.19, we can find several similarities with the Unity interface seen in Figure 6.12. It is worth noting that both engines allow their layouts to be customized by adding and removing panels, or by moving and resizing them. In this figure, we created a layout similar to the one we used in Unity to facilitate the comparison of the two engines. However, designers will often modify these layouts to prioritize interface elements that are important for their workflow.

The most prominent element in this layout is the **Level Viewport**, which displays the virtual scene (or "map," as it's called in Unreal Engine). The Level Viewport can be navigated with mouse or keyboard input. It also offers various tools for selecting, moving, rotating, and scaling objects while designing a map. The panel to the left of the Level Viewport is the **Outliner**, which shows a hierarchical list of the elements (or "actors") that populate the map. On the right-hand side is the **Details** panel, which displays the

Figure 6.19 Unreal Engine interface. The landscape, lighting, and player elements seen in the Outliner are included in the Unreal template project. An audio actor titled "Synth" has been added to the map.

properties of the actor currently selected in the Level Viewport or the Outliner. At the bottom is the **Content Browser**, which provides access to project assets, such as models, textures, and sounds. Another tab in this bottom panel is the **Output Log**, which is a debugging tool that displays errors, warnings, and printed messages during runtime.

6.8.1 Audio Actors in Unreal Engine

Much like in Unity, a map in Unreal Engine can be populated with various multimedia content, such as models, lights, sounds, and static or dynamic textures. In the following section, our second case study will explore some of the multimodal capabilities of Unreal Engine. For now, let's focus on its audio features.

Adding a sound source to a map in Unreal Engine can be done in several ways. In the Details panel of an existing actor, we can click the Add button and select Audio from the drop-down list to attach an **Audio Component** to the actor. When this component is selected, the Details panel displays a Sound tab where we can assign an audio file, a Sound Cue, or a MetaSound to the actor. In Figure 6.19, you can see an audio file titled "Synth" attached to the Audio Component of an actor. While this enables simple audio playback from the actor, we can define a more elaborate audio behavior using Sound Cues. A **Sound Cue** is a type of Blueprint that offers Sound Nodes, such as wave players, attenuators, crossfaders, Doppler and delay effects, and mixers. With a Sound Cue, we can implement a node graph that plays and manipulates audio files based on game state.

A Sound Cue file can be created by right-clicking in the Content Browser and selecting the relevant asset in the Audio submenu. Much like an audio file, a single Sound Cue can be attached to multiple Audio Components.

Although Sound Cues can be useful for implementing reactive audio features, Unreal Engine has introduced a more advanced procedural audio system that is likely to supersede Sound Cues in the near future. Released in 2022, **MetaSounds** is an audio programming platform built into Unreal Engine for designing custom audio graphs. It offers sample-accurate timing, which enables the creation of complex digital signal processing chains involving synthesizers, filters, and audio effects. Like Sound Cues, it interfaces with the rest of the engine to generate or modify sounds in real-time based on game state. Leveraging the computer technology advances that we touched upon in Section 6.3, MetaSounds enables the design of interactive audio graphs that would have been too computationally intensive in the earlier days of digital audio. As a result, it represents a promising evolution in game audio by facilitating a tighter integration between the audio, visual, and interactive modalities of a game.

MetaSounds offers a Blueprint-like visual scripting interface, where we can build audio graphs from a wide selection of nodes such as oscillators, filters, mathematical operators, and musical event controllers. This workflow is reminiscent of popular audio programming platforms like Max and Pd. In Section 6.7.2.1, we explored how Pd can be used for designing procedural audio in Unity. While this entailed a relatively straightforward workflow, the fact that MetaSounds is built into Unreal Engine allows for a more seamless interaction between game events and sound generation.

Besides fixed and procedural audio actors, Unreal Engine offers an **Audio Volume** actor, which is similar to the Audio Reverb Zone in Unity. Clicking the Create button in the Main Toolbar brings up the dropdown menu seen in Figure 6.20. The Audio Volume

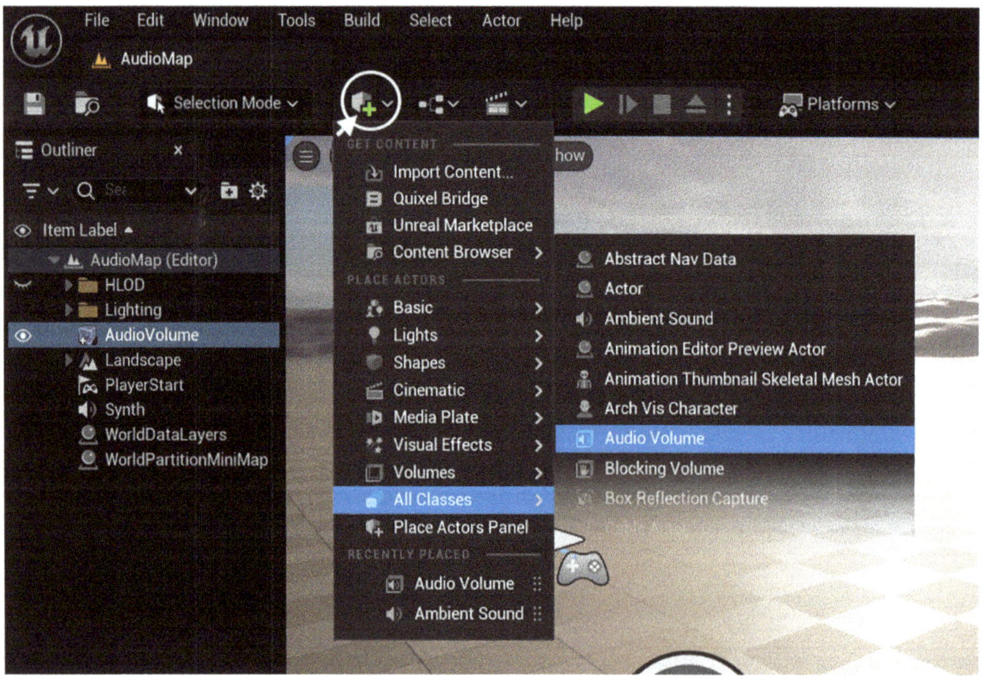

Figure 6.20 Drop-down menu to add an Audio Volume actor in Unreal Engine.

is listed under both the Volumes and All Classes tabs in this menu. Once a volume is added to the level, we can change its transform, size, and shape using the parameter controls in its Details Panel seen on the right-hand side of Figure 6.21. In this figure, we can also see the shape of the volume represented with a cube mesh in the Level Viewport. The Brush Shape parameter controls the Audio Volume's shape, which can be selected from a range of geometries including cube, cone, and cylinder, among others. We can then create a Reverb Effect asset as shown in Figure 6.21 and attach it to the Audio Volume. When a sound actor is placed within or moves through an Audio Volume, its sound can be fed into this Reverb Effect based on the actor's attenuation settings, which we will cover in the next section.

6.8.2 Attenuation and Spatialization

When a sound is added to a map, its Auto Activate parameter is enabled by default. As a result, the relevant audio file begins playing once the game is started. In the game, this sound is heard without any distance or localization cues (i.e., the user's movement

Figure 6.21 Right-clicking in the content browser panel brings up the pop-up menu seen here. The Reverb Effect asset is found under "Audio > Effects." This screenshot also shows the Details Panel for an Audio Volume actor. Here, a Reverb Effect can be attached to an Audio Volume. This is also where the volume shape can be changed using the Brush Shape parameter. Currently, this parameter is set to "Box," which is reflected in the Level Viewport.

through the virtual scene won't affect how the sound is heard). To introduce these cues, we would need to add a Sound Attenuation settings asset to the audio actor. This can be done by right-clicking in the Content Browser window and selecting "Audio > Sound Attenuation" in the pop-up menu. This will create a new Sound Attenuation settings file that can then be attached to one or more audio actors by dragging it to the Attenuation Settings box in the actor's Details Panel. This will allow the associated sound to be affected by distance and localization cues during gameplay based on settings defined in this file.

To change these settings, we can double click on the Sound Attenuation settings file, which will bring up the window seen in Figure 6.22. Here, we can set various properties affecting the spatialization of a sound. Looking at the Attenuation (Volume) tab, we see that sounds are given a spherical attenuation shape with a linear roll off by default, resulting in an omnidirectional propagation. This shape has an Inner Radius where the sound is heard at its full volume. The Falloff Distance indicates the range wherein attenuation is proportional to the player's distance to the sound source. The Inner Radius and Falloff Distance are visualized in the Viewport with meshes around the sound actor. When we change the attenuation shape or range, these visualizations are updated accordingly. The available attenuation shapes are capsule, sphere, box, and cone. Using the cone geometry, a sound can be made unidirectional. Much like the sound cone model used in Inviso, this shape will entail inner and outer cones for direct and cross-fading regions. The range and angle of the inner and outer cones can be controlled separately.

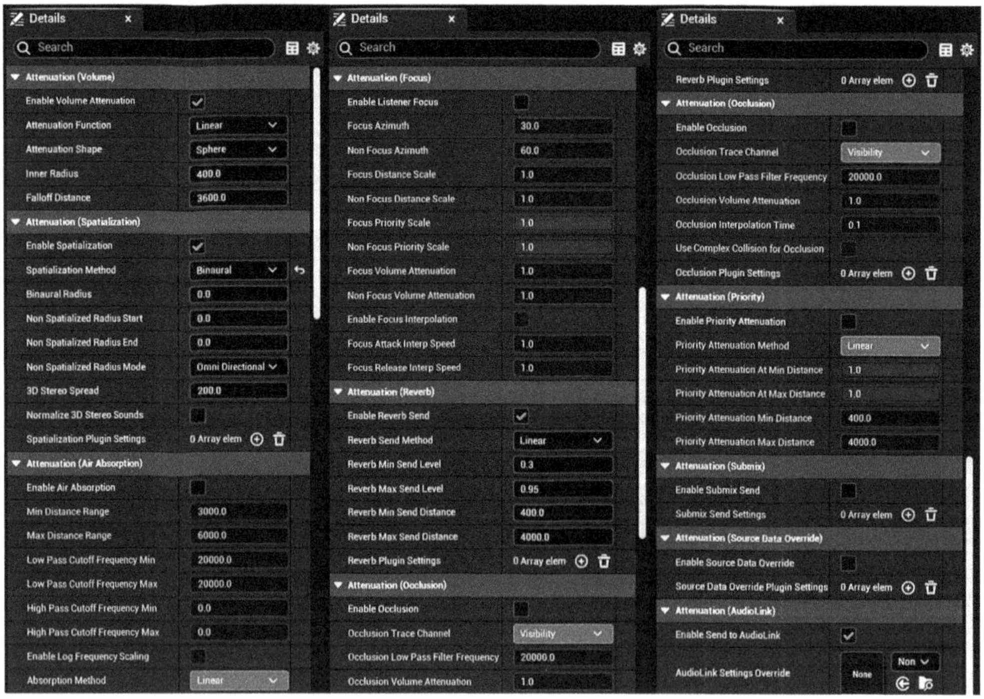

Figure 6.22 The parameters available in a Sound Attenuation settings file in Unreal Engine. Three screenshots from different portions of the settings are shown side by side.

Objects with sound cones can still be given an omnidirectional component with an additional falloff sphere.

The following tab in the Sound Attenuation settings gives us control over the Spatialization features of a sound. When the Spatialization Method is set to Panning, sounds are spatialized in stereo. If we prefer to set this to Binaural, we also need to specify a Spatialization Plugin under "Edit > Project Settings > Platforms > Current Platform > Audio." This can be set to a built-in or third-party plugin. The Binaural Radius allows us to set a maximum range wherein binaural rendering will be applied. The Non-Spatialized Radius Start and End parameters can be used to define a range around the sound actor where the source will not be rendered binaurally. This helps make a sound source take up the whole spatial field and not jump around drastically when the listener node is very close to it. The 3D Stereo Spread setting defines the spatial separation between the channels of a stereo audio file, allowing it to be perceived as a wider or narrower object in the scene.

The Air Absorption settings in the following tab help emulate the frequency dependent attenuation of sounds as they travel through the air. Since high frequencies decay more quickly than low frequencies over distance, this is essentially a low-pass filter applied to the audio actor to emulate an air absorption effect. We can set the distance range wherein this effect is applied. The Low-pass Cutoff Frequency settings define the cutoff frequency at the minimum and maximum distance settings. If a sound falls within this distance range, it will be affected by a low-pass filter whose cutoff is interpolated between the frequencies defined here. There is also a high-pass filter that can be applied to sounds within the absorption distance range. This can be used to attenuate low frequencies to diminish the perceived impact and size of objects at a distance.

The Listener Focus defined in the following tab is analogous to sound cones. This, however, controls the directivity of the listener node instead. By applying listener focus settings to a sound actor, we can give it greater prominence, for instance, when the player is looking directly at it. Note that although the focus settings seem to be applied to the listener node, they only affect the sound actors that have the Sound Attenuation file where these settings are defined. This also means that different sound actors can have different Listener Focus settings.

In the following tab, we can set both how much and within which range an actor's sound is passed to a reverb applied either globally through World Settings or, more likely, through an Audio Volume that has a reverb effect. Using the minimum and maximum distance settings, along with the corresponding minimum and maximum send levels, the degree of reverberation applied to a sound becomes a function of the distance between the sound actor and the player. This can be used to make nearby sound actors have less reverb than distant ones.

Although the Sound Attenuation file offers additional settings, the last item that we will cover here is Occlusion, which allows the sound propagation from an actor to be occluded by objects in the map before it reaches the listener. By default, occlusion is based on Visibility (i.e., whether a sound actor is visible to the player). However, the Trace Channel can be switched to a Collision Preset to allow only certain objects in the map to cause acoustic occlusion. We can set the cutoff frequency of the low-pass filter and the volume attenuation applied to the sound actor when it's occluded. We can also set how quickly the filtering fades in and out when occlusion starts and ends. The "Use Complex Collision for Occlusion" toggle determines whether the occlusion is based on the exact geometry of an obstructing actor or a simpler primitive geometry that bounds the actor, with the latter option being less computationally expensive.

6.9 Case Study 2: *Proximal*

In this section, we will present another case study on game audio in VR, this time focusing on Unreal Engine. This will allow us to not only understand the differences and similarities between the two major game engines used for VR development today but also explore Unreal Engine's procedural audio system MetaSounds. To this end, we will discuss the design and implementation of another VR music game, named *Proximal*, where the player is presented with eight spherical objects that emit sounds that are synthesized and spatialized in real time. While four of these play drum sounds (i.e., kick, snare, tom, hi-hat), the other four are synth instances that amount to four-voice polyphony. Using the game mechanics that we will describe in the following section, the player can create interesting rhythmic and melodic patterns. In addition to the sound objects, the player can interact with a control panel to modify the volume, BPM, musical scale, and decay time of the sound objects, as well as the gravitational force that affects them. The project files for this case study can be accessed at vraudio.org under Chapter 6. The reader is invited to download the project and explore its implementation in greater detail as they read through this section.

6.9.1 *Game Mechanics*

The core mechanic of the game is based on the player's distance to sound objects, hence the name *Proximal*. This distance is proportional to the rhythmic division at which an object's sound is triggered. In other words, when a sound object moves closer to the user it begins to produce sounds at a faster rhythm. As the player and objects move around in 3D space, the changes in the player's proximity to objects result in constantly evolving polyrhythmic patterns. Additionally, with objects that emit synth sounds, the distance is inversely proportional to the pitch, which is quantized on a user-selectable musical scale. Therefore, the synth objects play not just polyrhythmic but also contrapuntal melodies as they roll or bounce around.

The player can navigate the cylindrical room shown in Figure 6.23 either in room scale or through virtual locomotion using the hand-held controllers. The room contains half-tube slides on which the sound objects can spiral around the player. These slides weave in and out of the cylindrical room, making the sound objects move closer to and away from the center as they slide downwards. Since the pitch of a ball is inversely proportional to the distance between the ball and the user, it will follow an ascending pattern in its downward spiral around the vertical axis. However, the weaving of the slide will cause the ball's pitch to fluctuate around this pattern as it moves towards and away from the player in the horizontal plane.

Furthermore, portal rings that can be moved around allow the player to teleport objects in virtual space. The portals are color-coded to indicate how they are paired. When a sound object is thrown into a portal, it comes out from the corresponding portal. Using this mechanic, the player can teleport sound objects to different parts of the room. They can also vertically align portals to create teleportation loops.

Figure 6.23 A screenshot from *Proximal* designed in Unreal Engine. The image shows the cylindrical room wherein the sound objects can be set in motion in various ways.

Finally, using the control panel shown in Figure 6.24, the player can control global parameters pertaining to rhythm, musical scale, volume, and decay time. They can also adjust the gravity, introducing a degree of unpredictability to the objects' movement. The gravity slider works in a bipolar fashion—when it's set halfway, it's at the default gravity amount set by the engine. When pulled down, a downwards gravitational scalar is applied, making the objects less bouncy. When the slider is pulled up, an upwards gravitational scalar is applied. At higher settings, this causes the sound objects to levitate.

By throwing the sound objects around, setting them in motion on slides, teleporting them with portals, and by subjecting them to gravitational forces, the player can create complex polyrhythmic and polyphonic musical structures as they move around in virtual space.

6.9.2 Audio Implementation

The sound of *Proximal* is procedurally generated and spatialized in real time using the built-in features of Unreal Engine. Here, we will detail our implementation of synth and drum voices that make up the sound world of this game.

6.9.2.1 Procedural Audio

Much like in our first case study, the synth voices in *Proximal* are based on a square-wave subtractive synthesizer. MetaSounds offers built-in nodes for all the components of this type of synthesizer, including a square-wave oscillator with frequency and pulse-width modulation, a biquad filter for low-pass filtering, two and four-stage envelopes, as well

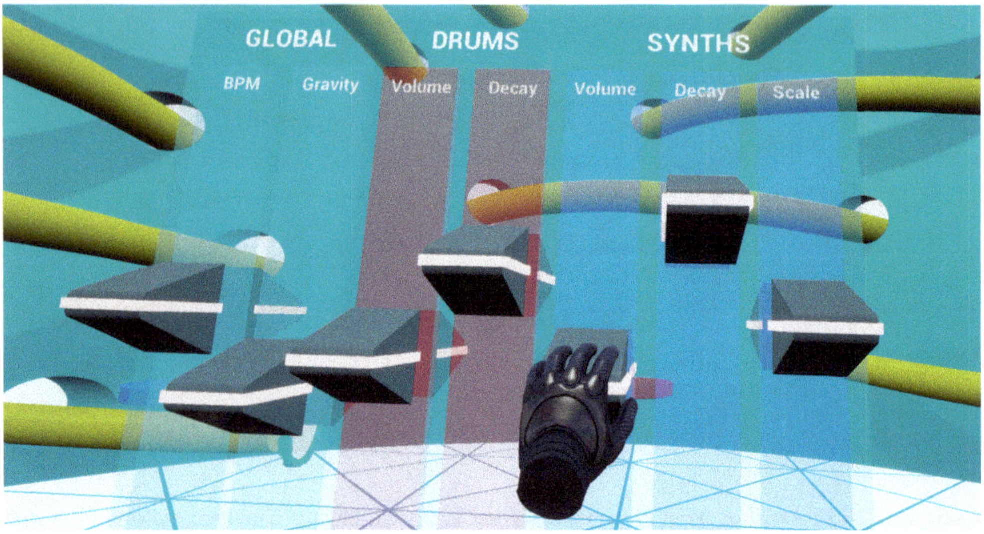

Figure 6.24 A close up of the control panel from the player's perspective in VR. The panel affords control over BPM, volume, and decay properties of the sound objects, as well as the scale of the gravitational force applied to them.

as math operators that allow audio signals to be mixed and amplified. In addition to audio processors, MetaSounds provides nodes for mapping and quantizing MIDI data.

The main musical clock is implemented with Unreal Engine's Quartz System in the scene's main event graph. Based on the BPM value set by the player, a clock signal is sent from the main event graph to the sound object Blueprints. This is then passed to the MetaSound graph associated with each sound object. Within this graph, a counter that is driven by the incoming clock and a scaled measure of the player's distance from the sound object are used as operands of a modulo operation. By selecting the instances where the modulo returns a zero, a subdivision of the global beat is derived for the sound object based on the player's proximity to it. This way, as an object moves further away from the player, it is triggered at a slower beat while still adhering to the rhythmic grid. The gain of the resulting sound is controlled by a global volume fader in the virtual control panel in the game, allowing the player to attenuate the sound. The output of this gain stage is sent to both the main output and an envelope follower that controls the brightness of a light source attached to the sound object. This complements the sound with visual feedback, where the object emits a pulsating light alongside its sound.

The architecture of a single synth voice in *Proximal* is illustrated in Figure 6.25. The flow graph in this figure is an abstraction of the Synth Voice MetaSound included in the project files. The figure also shows how the player distance and control panel parameters (i.e., Clock (BPM), Synth Volume, Synth Decay Time, and Musical Scale) are mapped to various parts of the synth. Although all four synth voices use the same MetaSound graph, each voice has a unique phase offset for the square wave, a sine-wave low-frequency oscillator (LFO) with a randomly-selected rate for frequency modulation (FM), and a

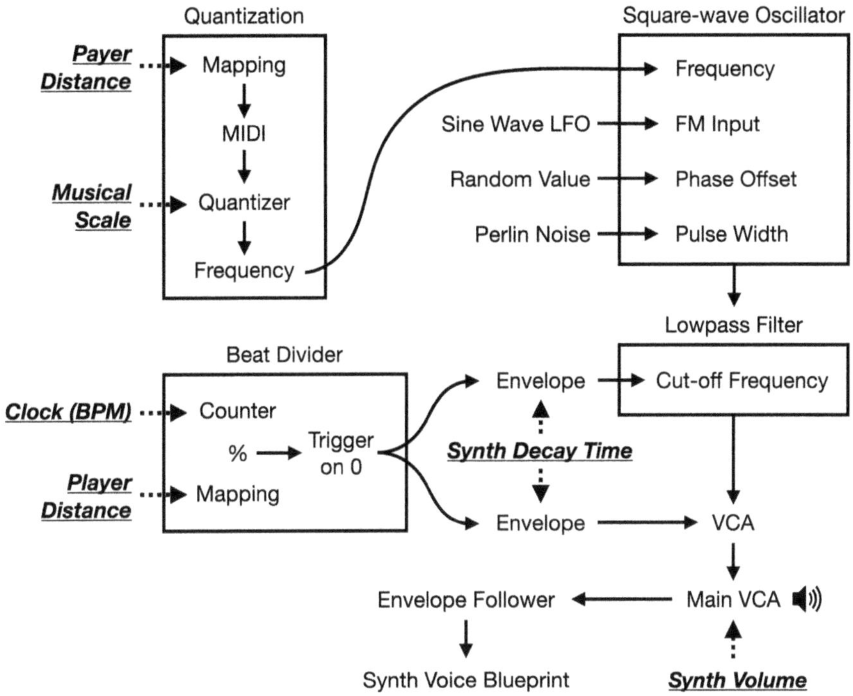

Figure 6.25 The architecture of the synth voice in *Proximal*. The underlined text shows the gameplay variables that control various aspects of the audio graph. Here, gain staging nodes are indicated with the VCA abbreviation, which stands for voltage-controlled amplifier. In electronics, VCAs allow a control signal to modulate the gain of another signal. In this case, a volume envelope gives a temporal contour to the output of the synth voice, whereas a slider in the virtual control panel allows the player to adjust the overall volume of the synth.

Perlin noise[9] source modulating the pulse-width of the square wave. Together, these create subtle timbral variations between the voices.

The design of the core drum sound is based on the flow graph displayed in Figure 6.26, which is an abstraction of the Drum Voice MetaSound included in the project files. The core drum sound is based on a combination of transient and resonance portions, which are marked with dashed boxes in the figure. By modifying these portions, different drum parts are achieved. For instance, the kick sound combines a transient sound consisting of a short noise burst with a longer resonance portion modulated by a pitch envelope, whereas the hi-hat is made up of a filtered noise burst with no resonance.

6.9.2.2 *Spatialization*

The sound of each object is spatialized binaurally using the built-in attenuation features of Unreal Engine. For the binaural rendering, we set the Spatialization Plugin under "Edit > Project Settings > Platforms > Current Platform > Audio" to Resonance Audio,

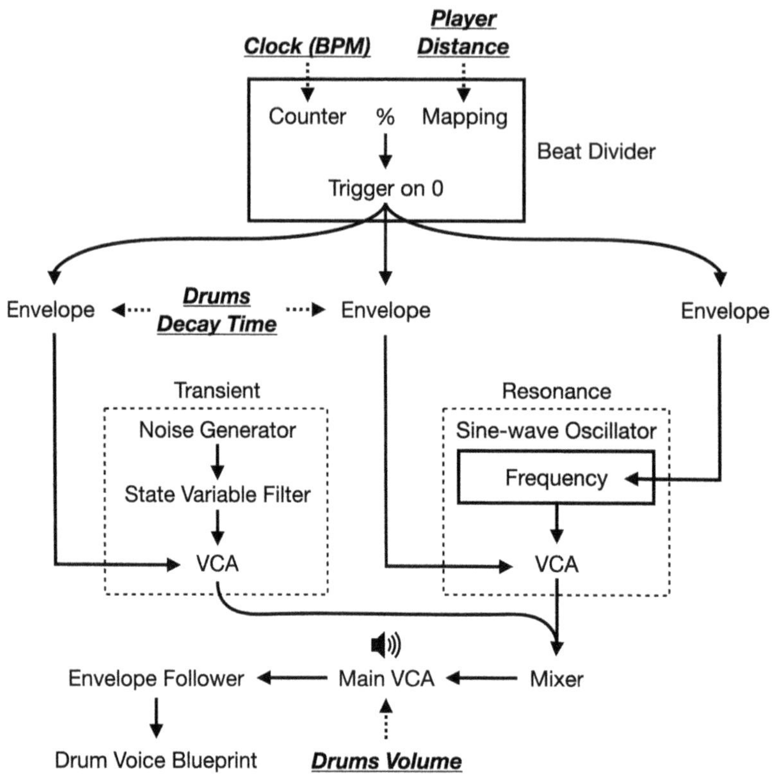

Figure 6.26 The architecture of the drum voice in *Proximal*. The underlined text shows the gameplay variables that control various aspects of the audio graph. While an envelope gives an amplitude contour to a filtered noise source for a transient sound, two other envelopes control the frequency and volume of a sine-wave oscillator for the resonance portion of the drum sound. Through different configurations of these two portions, we can get kick, tom, snare, and hi-hat sounds from the same MetaSounds graph.

which currently comes bundled with Unreal Engine. A Sound Attenuation settings file attached to each sound object defines directivity, range, and occlusion properties. Given their visual shapes, the sound objects are given a spherical attenuation pattern. Since all sound events transpire within the confines of the cylindrical room, we wanted to avoid overwhelming the player with too many loud sounds playing at once. To this end, we set the attenuation to a logarithmic function with a relatively short inner radius and a wider fall-off distance. Therefore, the objects are audible at a distance, but they reach their full volume only when they are close to the player.

To give the sound objects a shared auditory context, we added a reverb volume to the room. The sound of each object is sent to this global reverb based on the Sound Attenuation settings. Here, we set the Reverb Minimum Send Distance value to match the inner radius of the volume attenuation function. As a result, objects close to the player are heard without attenuation or reverb. As objects move away, their sounds are gradually attenuated and sent to the global reverb. The reverb send range extends beyond the attenuation fall-off distance, allowing fully attenuated objects to still contribute to the overall reverberant sound.

The spiraling slides cut in and out of the cylindrical room, causing the objects on them to occasionally go behind the walls of the room. When an object moves out of the room, its sound becomes occluded by the room's wall. The resulting lowpass filtering and attenuation of its sound gives the object an evolving timbre as it slides down.

Blueprints

In this project, we utilized Blueprints to implement game mechanics and link user interactions to procedural audio. While a flow-based visual programming environment like Blueprints can be more accessible than text-based programming languages, it still relies on programming fundamentals, such as variables, functions, and control statements. As with the Unity scripts in our previous case study, we included detailed comments in the Blueprints and MetaSounds that we designed for *Proximal*. Follow along these comments to gain a better understanding of how various parts of the game are connected to each other, even if you don't have prior experience with programming. Moreover, we advise leveraging these case studies as a foray into procedural audio programming. Proficiency in audio programming environments like MetaSounds and Pd can be an invaluable asset for working in VR and game audio.

In Chapter 1, we highlighted the rapid evolution of VR technology. An inherent implication of this is that the tools and workflows that we utilized in these case studies may be updated—or outdated—by the time you read this. Alternative tools with different features and interfaces are already available. However, the concepts and techniques that underlie the workflows presented here should still help you navigate these features and interfaces, as well as those that might emerge in the future. The current workflows offer a snapshot of audio creativity in VR today. But, more importantly, they are meant to offer our readers actionable perspectives on how to seize opportunities and tackle challenges in VR audio when dealing with both fixed and interactive media. In the next chapter, we will summarize these perspectives to paint a forward-looking picture of this field.

Notes

1 In the Dolby Atmos Renderer, the first ten input channels are reserved for bed tracks routed to fixed output channels, whereas Channels 11 through 128 are used for objects that can be moved in space.
2 https://valvesoftware.github.io/steam-audio/.
3 Unity also offers a visual scripting API that can be used for building game logic, although this is much less commonly utilized among Unity developers.
4 https://puredata.info/.

5 An example of this is *Spore*, which was developed by Maxis and published by Electronic Arts in 2008. The game allows the player to create unique species from microorganisms with visuals and music generated procedurally during gameplay. The procedural music is implemented in Pure Data with creative direction from the generative music luminary Brian Eno.
6 https://github.com/LibPdIntegration/LibPdIntegration.
7 https://learn.unity.com/course/beginner-scripting.
8 This is utilized as an alternative to the more traditional green screen technique, where one or more green screens serve as backdrops on set to be computationally replaced with images during post-production.
9 A type of gradient noise that can be used to modulate parameters in a way that is random yet smooth.

Chapter 7

Moving Forward

Guidance for VR Audio Practitioners

Over the course of the twentieth century, studies in audio and film production have fostered many theories and guidelines for practitioners. Similar frameworks have been proposed for video game design as well. Since VR can be viewed as a descendent of these more mature media fields, the best practices established for audio, film, and game production can inform emerging practices in VR. However, despite drawing on tools and techniques from these fields, VR brings forth conceptual and technical idiosyncrasies that require new perspectives on content creation. Therefore, we will need to adapt and reimagine some of the best practices developed for audiovisual media, on the one hand, and devise new ones that are tailored to the immersive medium, on the other.

In this chapter, we provide an overview of current best practices in VR audio. Some of these are rooted in those traditional media practices as well as the fundamentals of acoustics and psychoacoustics. As we discuss these guidelines, we take a human-centered approach, accounting for both the experience of the designer who works with immersive media tools and that of the user who engages with the outcome of that work. We also share some of the current industry perspectives on VR audio. Finally, we identify new directions for VR audio that we expect will offer unique creative opportunities for designers moving forward.

7.1 Towards a Set of Best Practices

In our effort to think through a set of best practices for VR audio, we start with conceptual recommendations conceived to stir our thinking around the role of sound in immersive media and more broadly, the role of immersive media in entertainment. We then move on to practical guidelines that summarize key insights from the case studies presented in the previous chapters.

7.1.1 Pursue Both Adaptive and Disruptive Approaches

When working in an emerging medium, creators tend to rely on the wisdom and knowhow from preceding media that are closely associated with it. Much like how early practices in television broadcasting adopted concepts and techniques from radio, early film work relied on prior knowledge in theatre and photography. As we underscored in Chapters 5 and 6, current practices in VR derive many tools and techniques from film, game, and audio production. In the case studies we presented, we utilized audio and film

DOI: 10.4324/9781003048022-7

editing software and game engines. Furthermore, we adopted some of the existing best practices in these fields to prepare, produce, and post-produce our projects. However, we also highlighted areas where VR required new approaches that are purpose-built for it as a creative medium. Therein lies the conflict and collaboration between adaptive and disruptive approaches to immersive media.

As we collectively identify a new set of guidelines for creating content for VR, it is natural for us to draw from our experiences in traditional media. These experiences can guide us through an uncharted territory of creativity and help us figure out the idiosyncrasies of the immersive medium. However, an adaptation of the knowledge gained from these experiences can go only so far in exploiting the unique possibilities of VR. It took a while for the television set to transition from "a radio with images" into a novel platform that capitalizes on the coupling between sound and images in news, entertainment, and sports. Similarly, VR will quickly grow beyond "a television with a dynamic first-person view."

To this end, we will need to disrupt some of our existing notions of content creation in audio, film, and games, and blur the distinctions between them to formulate new creative practices that are native to the immersive medium. Instead of creating gamified films, filmic music videos, or musical games in VR, designers will begin to craft realities that are greater than the sum of their parts. These will be participatory realities that provide users with a higher degree of agency and new modes of interaction. These realities will become increasingly social and promote multi-user and collaborative experiences.

Current guidelines for VR design often focus on realism, computational efficiency, and comfort. However, as new immersive media technologies and creative practices continue to emerge, and as computational resources become more abundant, some of these guidelines may become outdated or disproven. While following existing best practices can help designers overcome creative challenges, subverting these practices can lead to innovative approaches that might not be discovered otherwise.

We find an example of this in the early years of electronic music, where musical conventions helped composers navigate an entirely new creative medium. But once these composers became comfortable with electronic music technology, they began to exploit its novel affordances with new techniques that do not adhere to prior musical traditions (Çamcı 2022). Today, electronic music practices continue to push the boundaries of musical creativity. Similarly, many audio recording techniques that are common today originated from ill-advised studio experiments or outright accidents. In a famous example during the recording of Peter Gabriel's eponymous album in the late 1970s, the engineer Hugh Padgham was working with a new SSL console where a dedicated talkback microphone was routed into a heavy compressor and noise gate. The compressor made the musicians talking back to the control room in between takes uniformly audible, whereas the noise gate allowed this feed to be cut off when no one was speaking. When Padgham pressed the talkback button while Phil Collins happened to be playing the drums, the compressor accentuated the reverb in the live room, but the noise gate cut off the reverb tail as soon as Collins stopped playing. This led to the discovery of the famous "gated reverb" effect that finds common use in music production to this day. Although the effect entirely subverts the notion of realistic room acoustics, it gives the drums a distinct and full sound that blends particularly well with electronic sounds.

As much as we lean on proven methods in existing media practices, we should also strive to experiment and, in some cases, make mistakes. Balancing adaptive and disruptive approaches will ultimately lead to discoveries in workflow and creativity that are uniquely applicable to immersive media.

7.1.2 *Stay Agile as a Creator*

A critical component of working with VR is to recognize that it's an emerging medium—to the point that it may be called something entirely different in the future. In Chapter 1, we described how the distinctions between virtual, augmented, and mixed realities have increasingly blurred over the past few years. In addition to the umbrella term Extended Reality (XR), which is now being used to describe immersive media systems and experiences, Apple has recently popularized the term "Spatial Computing" for their Vision platform, which is by and large a VR headset in its current form. As VR technology rapidly evolves, so do the concepts and techniques that underlie immersive media practices. This is reflected in the frequent updates to software platforms and libraries that we use for VR design. While updating a piece of software can be a relatively straightforward process, more substantial paradigm shifts force us to rethink our approach to immersive media. For instance, the introduction of inside-out tracking and hand tracking methods in modern VR systems has prompted us to reconsider how we superimpose virtual spaces onto physical ones and how we interact with them. Therefore, it is crucial to remain agile and open-minded as we adapt to new hardware and software technologies, which, in return, bring about new creative possibilities.

At the same time, it is essential to consider the constant aspects of immersive experiences when working on a design. As much as VR may allow us to tamper with the laws of physics, the propagation of sound waves and, more importantly, our auditory perception adhere to principles that won't change with new technology. The dynamic and spectral thresholds of our hearing and the way we localize sounds in space are factors that will always govern our designs. While some of the tools and techniques that we discuss in this book may become obsolete, the perceptual underpinnings of these discussions will remain relevant. Therefore, it is advisable to refer to principles of acoustics and human perception when faced with a design challenge that eludes current technical or aesthetic frameworks.

7.1.3 *Reimagine the Role of Space in Music*

In the very early days of music production, all the parts of a song were recorded live onto a single track. Even when new recording technology allowed sounds to be captured into multiple channels, these had to be mixed down to one track because consumer playback systems only supported monophonic sound. When stereo sound became commercially available, the number of available tracks to record onto were still very limited. Moreover, the mixing consoles only offered left-center-right switches rather than continuous panning, so the instruments would often be hard-panned to the left or right channels. Once these technical limitations were overcome, it took a while for the music industry to fully embrace the potential of stereo, which prompted musicians and producers to conceive new ways to not only deliver but also compose their music.

In addition to the temporal, dynamic, and harmonic relationships between the instruments, music makers would now have to consider the grouping and separation of these instruments on a stereo soundstage.

Just as the switch from mono to stereo transformed the music industry, the undergoing transition from stereo to spatial audio will bring about new notions of music production. While the use of space is already a crucial component of music production, it becomes even more critical when the listener gains the agency to move both their head and their body in that space. Furthermore, spatial audio lends the engineer greater control over crafting and animating space. With stereo mixing, the engineer employs leveling, EQ, panning, and reverberation to construct a sound stage that offers both width and depth. This way, a stereo mix can envelop the listener and offer an immersive listening experience. However, mixing in spatial audio gives the engineer explicit control over the 3D positioning and movement of sounds. Instruments can be panned not just left and right, but also front and back, up and down, and everything in between. This enables the listener to be precisely situated within a 3D sound world, rather than simply being enveloped by a mix.

While the practical implications of spatial audio for music production can be easily extrapolated from stereo mixing, its aesthetic and conceptual underpinnings are more profound than a mere expansion of spatial control. In traditional mixing, we utilize EQ to give "height" to a track even though we don't have control over the elevation of a sound source. Now that spatial audio offers this control, we are left to imagine what parts of a song would merit a placement above or below the listener. For instance, a synth sound intended to add airiness to a track could be placed above the listener, whereas a bass drum could be below them towards the center.

Mixing engineers often use changes in panning and reverberation to create transitions between different sections of a song or to highlight various layers in a mix. While there is no inherent hierarchy between the left and right sides of a stereo panorama, there is a clear distinction between placing a sound source in front of or behind the listener. Common sense would suggest that placing a sound behind the listener might deprioritize its place in the mix. However, this could also evoke a sense of curiosity, drawing deliberate attention to that sound.

Spatial audio empowers not just the engineer but also the listener, who can now turn or move their head within a mix and observe it from different vantage points. This interactivity makes the spatial design of a song even more crucial, as it invites the listener to actively engage with the mix. The listener's interaction with space brings about new narrative and compositional possibilities. In popular music, it's fairly common for a song to tell a story. The various phases of the production process from arrangement and orchestration to recording and mixing serve to deliver this story. Spatial audio adds a new dimension to this process by allowing the elements of a song to occupy and move within the 3D space around the listener. These can cue the listener to move their attention within the mix. Some of these elements can be veiled behind spatial layers to be discovered through the active participation of the listener. As mixing gains a more pronounced narrative function, artists can envision their music as an interactive experience and compose its space alongside its rhythmic, timbral, harmonic, and lyrical dimensions.

The idea of composing musical spaces through spatial audio has been extensively explored in electroacoustic music throughout the twentieth century (Vaggione 2001; Barrett 2002). However, these explorations have generally relied on complex speaker setups that the average consumer doesn't have access to. Spatial audio has been largely exclusive to academic research with design tools that are catered towards experts in the field. With the greater availability of VR and spatial audio systems in the past decade, some of these design tools, such as the free software that we demonstrated in the previous chapters, are becoming more easily accessible. Today, it's much easier for an independent artist to pick up a DAW plugin that allows them to design an Ambisonic space for VR. Therefore, the use of spatial audio is quickly expanding from the confines of academic research into the workflows of professionals and enthusiasts alike.

Currently, the music industry is experiencing a major trend where existing records are being adapted from stereo to spatial audio. While some of these adaptations work reasonably well, others fall short of doing justice to the original material, which wasn't conceived with spatial audio in mind to begin with. Moving forward, we can expect the entire creative personnel involved in the making of a track from composers and performers to producers and engineers to contribute to the vision behind the use of space in a track. With works that are native to the immersive medium (i.e., those that are not merely adapted to but altogether devised for immersive delivery), spatial audio will live up to its full promise in music production.

Beyond its impact on spatial audio and, by extension, on music production, XR technology is also reshaping the relationship of musical creativity with interactive and visual media (Park, Choi, and Lee 2024). An example of this is music concerts delivered to audiences in VR. These not only increase the public's access to immersive concert experiences but also offer unique perspectives into music performance and opportunities for audience participation. We showcased an example of this in Section 5.6 with a performance that situates the viewer in the middle of a percussion ensemble. Mainstream artists like Doja Cat, Megan Thee Stallion, and Paul McCartney have similarly utilized VR to bring audiences to the stage. Such experiences also introduce a new degree of interactivity, where the user can change where they are situated and even alter elements of the music.

An application of VR concerts besides entertainment is in music education and musical training. An increasing number of educators and practitioners are using VR for exposure training, where music students can perform in various concert spaces to virtual audiences or professional juries (Bissonnette et al. 2016; Bellinger et al. 2023). These applications have shown promise in helping with stage anxiety, which can severely impact the performance and mental health of musicians.

7.1.4 *Justify the Use of the Immersive Medium*

It goes without saying that the immersive medium is not appropriate for all media content. Many films and games are better presented on a 2D screen, and many interactive applications are better controlled with traditional input devices such as a keyboard or mouse. Similarly, stereo will continue to be a prominent format for music delivery. When planning for VR content creation, a designer should carefully consider if VR is the best presentation format and avoid using it just for its novelty. It is a good exercise for the designer to ask themselves at every step of the creative process whether the idea at hand justifies the use of the immersive medium. Such an exercise isn't necessarily intended to identify reasons to abandon the use of VR in a project but rather to enhance the work's capacity to fully exploit the affordances of the medium.

We have already discussed some of these affordances in the previous chapters. Foremost among these are multimodality, magicality, and agency. In our daily lives, sound doesn't transpire in a sensory vacuum. It's often accompanied by visual or tactile stimuli stemming from the event that gives off the sound. Even when a sound is presented to us in isolation, our minds associate it with other sensory modalities (Warren, Kim, and Husney 1987; Taruffi and Küssner 2019). VR bolsters the multisensory contextualization of sound by not only pairing it up with stimuli in other modalities but also allowing us to perceive and interact with it in a way that closely approximates our relationship with sound in our daily lives. However, a faithful simulation of reality is hardly the end goal of a VR experience.

This brings us to the affordance of magicality. When the multimodality of VR is teamed up with the ability to distort reality, the experience becomes magical. We can bend the rules of physics and go to places unvisited or unimagined. Interactivity and magicality alone are not novel concepts in digital media. Games, films, and even software applications can be interactive and feel magical. However, VR couples these qualities with a first-person view and self-representation based on hand, body, and head tracking. This brings us to the affordance of agency in VR. In traditional computational systems and media experiences, our agency is often deferred. The keyboard relays our button presses to applications, the mouse cursor represents our hand, and the camera in a game or film serves as our eyes. VR brings all these interactions back to us; we see VR from the vantage point of our own eyes and we interact with magical worlds through virtual manifestations of our bodies.

7.1.5 Leverage Procedural Audio

When we view the intersection of multimodality, magicality, and agency in VR from a sound design perspective, one of the most promising trajectories is procedural audio. As we covered in Section 6.3, procedural audio involves the control and generation of sound and music during gameplay based on player interactions and game state. In our case studies, we presented examples of fully procedural audio, wherein all sounds are synthesized in real time within the engine. Procedural audio allows the designer to define a parametric relationship between sounds and visuals in a game, creating a tighter connection between the different modalities. Furthermore, procedural sound synthesis during game play expands the sonic scope of a game beyond a fixed set of pre-recorded sounds that are triggered based on game events. Audio synthesis in general offers an unlimited range of sounds, thereby creating opportunities for designing magical sounds that do not correspond to any real-world events. This allows not only each gameplay to have a different sound design or soundtrack but also each individual user action to have a unique sound to it.

This brings us to the impact of procedural audio on agency in VR. Whereas the benefits of procedural audio for multimodality and magicality are applicable to traditional games as well, the first-person view of the game world and the user's embodied interaction with the game through various tracking systems make procedural audio more critical in VR games. In traditional games, most of the user input is mapped to first- or third-person character controls that trigger predefined animations. However, user interactions in VR can be as embodied and intricate as our interactions with real-world objects. While it is plausible for a pre-determined character movement to have a consistent sound in a traditional game (e.g., the sound of unlocking a door), it can be immersion breaking if the same sound is played for different instances of a VR interaction. Procedural audio can allow such interactions to have dedicated sounds that match the exact nature and extent of the interaction. Furthermore, the addition of 6DoF spatial audio in VR gives the user a much more dynamic and detailed perspective towards sound and enables greater acuity in listening. Procedural audio can elevate the realism and responsiveness of sound-producing events under this expanded scrutiny of the user in VR.

7.1.6 Enhance VR with Spatial Audio (and Vice Versa)

Throughout this book, we have emphasized the critical role of sound in establishing presence and immersion in VR. Since the auditory phenomena we encounter in our daily lives are inherently spatial, it only makes sense to spatialize as many sound sources as computationally possible when designing a VR experience. This excludes instances where a non-spatialized delivery serves a narrative function, for instance, when non-diegetic voice-over is presented to the user without any spatial cues.

While the field of view in current VR headsets is relatively restricted, the field of hearing is unlimited, giving the user a peripheral awareness that can't be achieved with visuals alone. Therefore, using spatial audio to surround the user with sound sources serves a practical function when guiding their attention in VR. Besides a sound source's azimuth and elevation, its distance contributes to its perceived location. Although most spatial audio implementations in game engines have a default distance model, the designer can tune this model to strike a balance between realism, isolation, and auditability.

By default, most spatial audio instances are point sources, whether they are audio components attached to objects in a game engine or spatial audio channels in a DAW plug-in. However, as we discussed in Section 6.5.2, rarely any sound in real life propagates from its source in a uniformly spherical pattern. Especially in VR games, using directional sources can make the audio feel more focused and realistic. It can also prevent sound sources from interfering with each other as much as they might with an omnidirectional sound propagation pattern.

Speaking of realism, we talked about how it's not just the sound itself but also its relationship with its environment that governs how we perceive it in space. The designer should pay as much attention to the acoustics of the virtual space as they do to the placement and directivity of the sounds that propagate in that space. By using geometrical acoustics when computationally feasible or by fine-tuning an added reverb effect, the designer should ensure the reverberation in a virtual environment closely matches the features of the environment. Other room effects, such as air and surface absorption, should be utilized to give additional realism to the propagation of sounds based on the size and surface materials of the environment. Finally, occlusion effects are critical for a convincing relationship between the sound source, the virtual environment, and the user. This is another area where geometrical acoustics can prove useful for the designer.

On the other side of the relationship between spatial audio and VR, the head tracking found in modern VR systems offers distinct benefits for spatial audio. In Section 3.4, we discussed how head-related transfer functions (HRTFs), which represent a mathematical model of our perception of sound in space, are unique to each individual. However, we also described how it's impractical on a commercial scale to derive individualized HRTFs for each user. This is why most spatial audio implementations today rely on non-individualized HRTFs that are averaged from a limited segment of the population. The mismatch between a non-individualized HRTF and a user's anatomy leads to inaccuracies in spatial imaging that manifest as front-back and up-down confusions. Although we encounter such confusions in our daily lives as well, we use minute (and often subconscious) head movements to sample the auditory scene from slightly different perspectives and resolve conflicts in spatial perception. This is where VR offers a leg up for spatial audio. When head-locked binaural audio suffers from an HRTF mismatch, the only solution would be to try another non-individualized HRTF that is hopefully closer to the listener's anatomy. However, when head tracking is involved, the listener's head movements can help resolve some of the errors in imaging that stem from an HRTF mismatch. Sure enough, research shows that head tracking improves localization accuracy with non-individualized HRTFs (Su, Marui, and Kamekawa 2019).

7.1.7 Consider Audio Fundamentals

In Chapters 5 and 6, we identified several audio strategies for VR film and games. Following the adaptive approach that we identified earlier in this chapter, some of these were derived from traditional practices in audio production. It goes without saying that, regardless of the delivery medium, the designer must strive to achieve the highest sound quality possible. To that end, they should combine critical listening with a good understanding of acoustics and psychoacoustics. When the designer encounters a technical or aesthetic hurdle, trusting their ears, as well as their understanding of propagation and perception of sound, can help them come up with a solution.

In order to make their releases as compact as possible, game companies may prefer compressed or lower-quality sound files that take up less space at the expense of audio quality. Similarly, the audio quality in a film can be influenced by the standards adopted in delivery platforms. But even before addressing the artifacts that may arise from data and delivery formats, the designer should ensure the raw audio material is recorded, synthesized, or stored without unwanted distortions, pops, or clicks. In fact, the influential spatial audio researcher Durand Begault lists the elimination of fatiguing aspects of sound, such as noise and distortion, as a key requirement for high quality 3D sound (Begault 2000).

Low-quality audio files may suffer from a compressed dynamic range, perceptual encoding artifacts, and limited frequency bandwidth, none of which can be recovered during the design process if the raw material is already of low quality. That is, unless the designer prefers these artifacts for aesthetic purposes. However, even when a limited frequency spectrum is sought after for lo-fi aesthetics, the designer might need to deal with it differently in VR audio. Due to the physiology of our auditory system, we are better at localizing high-frequency sounds. This is why broad-spectrum sounds, particularly those that are not dampened in the higher range of the frequency spectrum, are critical for creating a lively and dynamic spatial scene around the user. The same characteristic of our auditory system can also be exploited for computational efficiency. Since we are not particularly good at localizing low-frequency sounds, these can be used without spending computational resources on their spatialization.

Another aspect of our auditory perception is that we are always hearing *something*. Even in the quietest environment, we can hear sounds if we listen closely enough. And if there is really no sound around us, we begin to hear the sounds of our own body, like the beating of our heart or the flow of our blood. In other words, we are never in absolute silence. This is why the designer should avoid total silence in VR audio. Because we expect at least a minimal level of sound in real-life environments, the lack of sound in a VR experience can break immersion. Even if there are no visible sound-producing events in the experience, the designer can use an Ambisonic room tone to instill a sense of presence.

One of the strictest guidelines for the visual aspect of VR is to maintain a high framerate to avoid motion blur and delayed head tracking, as these artifacts can cause simulation sickness and nausea. An auditory counterpart to this is audio latency (i.e., the delay between user input and its audible outcome). This issue becomes particularly problematic in interactive audio applications, such as games where user actions trigger sound events. Audio latency can occur for several reasons, some of which may be external to the audio implementation itself. For instance, if the intensive graphics processing in a game begins to affect frame rate, this can also impact audio performance and lead to latency. Similarly, if an audio file associated with a game event requires excessive buffering, it can delay playback. The procedural audio in a game may also suffer from latency if not implemented with efficient algorithms and audio processes.

The resulting latency, which can be jarring in traditional video games, can cause discomfort and simulation sickness in VR. Although the human mind can bridge minor temporal discrepancies between a visual element and the sound it produces, latencies that exceed a few hundred milliseconds between the two modalities can disrupt the sense of presence. It is therefore advisable for the designer to maintain low audio latency and a tight coupling between sound, visuals, and user input. That is, unless disassociating sounds from visuals functions as a narrative device or game mechanic, such as depicting a dream state in a game. Even then, the designer should ensure that the latency does not lead to any physical discomfort.

7.1.8 Prioritize User Experience

On that note, the VR audio designer should always mind the user experience. Earlier in this chapter, we encouraged designers to challenge themselves to justify the use of the immersive medium from a creative standpoint. Another consideration when weighing the use of VR is the level of commitment required from the user, especially given the limitations of current VR technology. Using modern VR systems continues to be a more involved process than casually picking up a phone or opening a laptop. It often involves finding an appropriate place to put on the headset and even clearing up a play area if the experience involves room-scale tracking. Moreover, current VR headsets are still relatively large and heavy, and they tend to get warm with extended use, making them uncomfortable for prolonged sessions. This further underscores the importance of making a VR experience worthwhile for the user by leveraging the capabilities of the immersive medium to the fullest.

Firstly, the designer should consider how long they expect their application to keep the user in VR. Different users will have different levels of tolerance for the physical strain that current VR systems can place on them. Additionally, certain applications will require the user to exert greater physical effort through bodily movements and even physical locomotion in room-scale environments. With these factors in mind, creating applications that can be experienced in short bursts or those that offer breaks between play periods will prevent users from becoming exhausted.

A specific type of exhaustion related to sound is called ear fatigue. This occurs when someone strains their hearing over a long enough period that they begin to experience side effects ranging from loss of sensitivity to physical discomfort. In the tail end of the human auditory pathway in the inner ear, tiny hair cells in the cochlea convert mechanical vibrations transmitted through the ear canal into electrical signals that the brain interprets as sound. These hair cells, much like muscles, can become overworked with extended use. In extreme cases, this can lead to permanent damage in the form of tinnitus and hearing loss. Two critical factors that contribute to ear fatigue—and, in more severe cases, hearing damage— are the intensity of exposure (how loud a sound is) and its duration (how long the listener is exposed to it). The designer should carefully consider both of these factors to avoid creating an experience that, at best, is aesthetically unpleasant and, at worst, poses a serious health risk. Note that most VR systems offer physical buttons for adjusting the audio output level. The designer should ensure that even at the highest volume settings the application does not cause any kind of temporary or permanent hearing damage.

Even when a VR application doesn't induce a physiological response like ear fatigue, it can still overwhelm the user cognitively. When experiencing a traditional film or game through non-immersive media, the user can easily distance themselves from a scene by looking away or even closing their eyes or ears. However, in VR, simply looking away from an object of interest may not provide the same level of detachment from the content. Given the emerging nature of VR today, a designer's work may be one of the first VR experiences for some users. In such cases, the user won't have any prior reference or guidance on how to navigate their experience, making them more susceptible to being overwhelmed than when they are viewing traditional media.

Many users who try VR for the first time experience a higher degree of simulation sickness than more experienced users. While visuals often play a greater role in this type of discomfort, the VR audio designer should also refrain from overwhelming the user. While moving sounds can make for an engaging experience in VR, too many sounds that constantly zip by the user can be disorienting. Even though the sounds themselves may not be uncomfortable to experience from a perceptual standpoint, the urge to follow them in space can induce discomfort. Keep in mind that more experienced users might fare better in such situations. However, if an application is intended for a general audience, the designer should consider the least experienced users as the baseline and either adjust their design accordingly or offer tiered designs that target different experience levels. The onus is on the designer to foresee the potential pitfalls in the experience they offer and address the ones that might cause discomfort for the user.

It can be better to build tension and climax by juxtaposing animated moments that have a lot of spatial audio sources with those that are calmer. This would both impart greater narrative significance on those animated moments and offer the user a moment to breathe in between them. These rests in the intensity of spatial audio can also help avoid ear fatigue.

Even when a spatial audio scene in VR is not overwhelming, overuse of spatial audio for sounds that are not meant to attract the user's attention can distract them from what they are expected to focus on in immersive space. This can adversely affect the narrative experience and detract from the intended progression of events in a game or film.

7.1.9 *Playtest Your Designs*

A reliable method to ensure a VR project is an enjoyable and comfortable experience for the target audience is to carry out playtests with users during the development phase. A common challenge in creative work is that the designer can gradually become too accustomed or desensitized to the project they are working on. This results in a loss of perspective, where the designer can unknowingly get used to the potentially uncomfortable aspects of their work. Gathering feedback from users at various stages of development can help address these blind spots. This can be as simple as asking a friend or a family member to try out the work and describe their experience. This will provide a fresh perspective on the work and help identify potential issues early on.

7.1.10 *Strategize about Sound Early On*

VR fosters a tighter integration of modalities, where sounds, visuals, and user interactions are aligned not only in time but also in space, under the heightened scrutiny of the user, who now has a dynamic point of view and additional degrees of freedom. As demonstrated in our case studies in Chapter 5, capturing sound for VR requires extra planning for equipment selection and placement, and the decisions made during the production phase significantly impact post-production. Similarly, our case studies in Chapter 6 demonstrated that the use of spatial and procedural audio in VR games and, more broadly, interactive experiences, requires early incorporation of audio design and programming into the development process. While advance planning is important for any media project, strategizing sound in VR is even more critical to ensure the production or development phases of a project proceed smoothly and the final outcome fulfills the potential of the immersive medium.

7.1.11 *Question the Impact of VR on Society*

We present this as a closing consideration not because it's an afterthought but more so to underscore its importance in the formulation of best practices in VR. Media platforms have an immense effect on both how content is created and how it's delivered and consumed. The platforms that we consider as "traditional," such as radio, television, and, more recently, the internet, have all caused societal paradigm shifts in terms of how populations inform, educate, and entertain themselves. On the one hand, these platforms accelerate our access to information, make us more interconnected, and empower our creative endeavors. On the other hand, they can be misused to spread disinformation and marginalize or exploit communities.

Although it can be exhilarating to explore the unique possibilities of VR for artistic expression, we should also maintain a critical perspective towards it and question its impact on society as we produce content for it. With an emerging medium like VR, predicting the influence of our work can be challenging, particularly in the absence of tried and tested guidelines like those we have for traditional media platforms. This places a greater responsibility on designers to think carefully about the ethical implications of their work.

Recent research has shown that the sense of presence and embodiment in VR can instill a feeling of connection with others and an understanding of their perspectives (Ventura et al. 2020). This is why VR has been described as an empathy machine. In a study on the role of VR in immersive journalism, researchers have shown that an article presented in the form of a VR film had a stronger effect on human rights attitudes compared to when it was presented as a non-immersive video or in a written format (Bujić et al. 2020). In another study, the VR version of a United Nations documentary on refugee camps was found to evoke greater engagement and a higher level of empathy than when it was presented in 2D (Schutte and Stilinović 2017). However, the capacity of VR to instill empathy can also make its users vulnerable to exploitation. Long before the recent wave of VR technology, the ethics researcher Philip Brey argued that the interactive and representational features of VR can help spread biases and misrepresentations (Brey 2020). More recently, researchers have also raised concerns about the potential effects on VR technology on children's psychosocial development (Kaimara, Oikonomou, and Deliyannis 2022).

Although VR has a long history in academic and artistic communities, the leading force behind it today are Big Tech companies. The significant investments that these companies make in the research and development of VR technology in the past decade have undoubtedly propelled the medium to unprecedented levels. These developments now empower the work of those academic and artistic communities. However, the vision that these companies promote for VR may not always align with the interests of artists, researchers, or even customers. Unsurprisingly, these technological developments and their commercial applications are aimed at generating profit, making the monetization of immersive experiences an inevitability. While this would be inherently beneficial for the growth of VR as a creative medium, the methods of monetization could introduce societal challenges and risks.

One such risk is the use of "deceptive patterns," which are interaction and information design techniques utilized in social media, games, and other online services to manipulate users into spending more time on a platform, making purchases, or relinquishing their privacy. VR introduces greater degrees of user interaction with head, hand, body, and gaze tracking, which enable VR systems to gather detailed information on the behavioral patterns of a user. Most modern VR systems also use inside-out tracking, which relies on scanning the user's living and working environments, raising further privacy concerns. Moreover, these systems can take up the entire audiovisual space around the user and capture their attention more effectively. This is an area where audio can be particularly powerful as it can easily envelop the user and influence how the physical and virtual worlds are bridged.

In a systematic analysis of deceptive design in XR, researchers have both found clear parallels with techniques used in existing media and identified new deceptive patterns specific to the immersive medium, such as those that exploit multi-sensory feedback to manipulate users (Hadan et al. 2024). While an individual content creator may not have much sway over the commercial strategy of a Big Tech company, a community of creators and their collective attitude can be consequential for the growth and evolution of a platform. As creators, it's natural to strive to make our work engaging, but we must also ensure the same work does not become exploitative. This is an area where our previous guidance on designing with user experience in mind and working with users during the design process can be particularly effective.

7.2 Two Industry Perspectives

We have already touched upon the influence that some of the Big Tech companies have on the evolution of VR. These companies steer the VR industry by not only building VR hardware and software but also by formulating use cases for their products and marketing them accordingly. They are also among the leading producers of content; many VR hardware manufacturers develop games, films, and other interactive media for their platforms, employing numerous researchers, developers, and artists.

Two companies who are leaders in the VR industry today are Meta and Apple. Meta (formerly known as Facebook) has been publishing VR best practices for several years now. In that time, the company has made numerous updates to these documents based on new hardware features (e.g., inside-out tracking, passthrough view), interaction schemes (e.g., hand-based input, virtual locomotion techniques), and usability research. Apple has recently introduced similar guidelines for their Vision Pro. While occasionally exclusive to the platform they are written for, the guidance provided in these documents are generally applicable to VR development in other platforms as well.

In this section, we outline some of the audio-related best practices established by these two companies. As we have noted before, these best practices will continue to be a work in progress in the near future while the VR industry matures. It is also likely that some of the best practices that we quote below will have been modified by the time you read this. Nevertheless, these give us a glimpse into industry perspectives on VR audio and its implementation. You may also recognize some similarities and differences between the recommendations laid out by industry leaders and the guidelines that we discussed throughout this book. Note that we abridge these best practices by discarding some of the general discussions on acoustics, auditory perception, and platform-specific features.

7.2.1 *Meta Quest*

From the "VR Audio Design, Engineering and Mastering Guide" published online:[1]

> While there is a lot that goes into delivering a high quality, immersive VR experience, audio is an element that is sometimes overlooked in regard to how essential it is in delivering that heightened sense of immersion ... Whether you're an aspiring sound designer or a veteran, VR provides many new challenges and inverts some of the common beliefs we've come to rely upon when creating music and sound cues for games and traditional media...

- Spatial audio is essential to deliver an immersive experience because it provides powerful cues that make the user feel they are actually in the 3D environment ...
- Headphone audio will be the standard for VR into the future, as it provides better isolation, privacy, portability, and spatialization... (With headphone audio) head tracking is greatly simplified... HRTFs are more accurate since they don't suffer from the "doubling down" of HRTF effects, i.e. sounds being modified from the simulated HRTF, and again by the listener's actual body geometry.
- Audio immersion is maximized when the listener is located inside the scene, as opposed to viewing it from afar ... an audioscape in which moving elements whiz past the listener's head with auditory verisimilitude is far more compelling than one in which audio cues cut the listener off from the action by communicating that they're outside of the field of activity.
- Most spatialization techniques model sound sources as infinitely small point sources; that is, sound is treated as if it were coming from a single point in space as

opposed to a large area, or a pair of discrete speakers. As a result, sounds should be authored as monophonic (single channel) sources.

- Broad spectrum sounds (such as noise, rushing water, wind sounds) spatialize very effectively providing lots of frequencies for the HRTF to work with. They also help mask audible glitches that result from dynamic changes to HRTFs, pan, and attenuation. In addition to a broad spectrum of frequencies, ensure that there is significant frequency content above 1500 Hz, since this is used heavily by humans for sound localization.
- Low frequency sounds are difficult for humans to locate—this is why home theater systems use a monophonic subwoofer channel. If a sound is predominantly low frequency (rumbles, drones, shakes, etc.), then you can avoid the overhead of spatialization and use pan/attenuation instead.
- When it comes to sound design for immersive experiences, it may come as a surprise, but realism is not necessarily the end goal ... This is especially true when considering issues such as dynamic range, attenuation curves, and direct time of arrival.
- It is important to position the sound emitter at the correct location on a character (e.g. footsteps from the feet, voices from the mouth) to avoid weird phenomena like "crotch steps" or "foot voices".
- Modeling time of arrival delay may paradoxically make things seem less realistic, because it introduces additional latency and can make it feel like the sound is out of sync with the visuals. Also keep in mind that we are conditioned by popular media to believe that loud distance actions are immediately audible.
- Stereo sounds will not be positioned in the virtual world and will not respond to head tracking. This makes the sounds appear "head locked," as they follow the users head movements rather than feeling grounded in the virtual world. This can detract from the spatial audio experience and should generally be avoided when possible ... For original compositions it's best to mix to ambisonics which can be rotated and won't be head locked. If that is not an option, then try to be mindful of how the music impacts the spatial audio.
- Performance is an important consideration for any real time application ... In general, you shouldn't be too limited by performance overhead of spatialization but it's important to know your audio performance budget and measure performance throughout development.
- While latency affects all aspects of VR, it is often viewed as a graphical issue. However, audio latency can be disruptive and immersion-breaking as well ... High system latency becomes an issue as the relative speed between an audio source and the listener's head increases. In a relatively static scene with a slow moving viewer, audio latency is harder to detect. As a ballpark, around 100ms is the threshold where the delay for head rotations is noticeable for most users.
- Effects such as filtering, equalization, distortion, flanging, and so on can be an important part of the virtual reality experience.
- When mixing scenes for VR ... properly spatialize sound sources; create soundscapes that are neither too dense nor too sparse; avoid user fatigue; use suitable volume levels comfortable for long-term listening; design with appropriate room and environmental effects ... Make sure important sounds are clearly heard even at a distance, and unimportant sounds don't clutter the mix ... When mixing for VR

there is an opportunity for heightened immersion if we provide the correct audio cues, so it's important to consider how attenuation curves impact perception of distance ... Controlling the amount of reverb per sound is a critical component to creating the perception of distance.

7.2.2 *Apple Vision Pro*

From the online design guidelines for "Playing Audio" on the visionOS:[2]

Subtle, expressive sounds are everywhere in visionOS, enhancing experiences and providing essential feedback when people look at a virtual object and use gestures to interact with it. The system combines audio algorithms with information about a person's physical surroundings to produce Spatial Audio, which is sound that people can perceive as coming from specific locations in space, not just from speakers...

Prefer playing sound. People generally choose to keep sounds audible while they're wearing the device, so an app that doesn't play sound—especially in an immersive moment—can feel lifeless and may even seem broken. Throughout the design process, look for opportunities to create meaningful sounds that aid navigation and help people understand the spatial qualities of your app.

Design custom sounds for custom UI elements. In general, a system-provided element plays sound to help people locate it and receive feedback when they interact with it. To help people interact with your custom elements, design sounds that provide feedback and enhance the spatial experience of your app.

Use Spatial Audio to create an intuitive, engaging experience. Because people can perceive Spatial Audio as coming from anywhere around them, it works especially well in a fully immersive context as a way to help an experience feel lifelike. Ambient audio provides pervasive sounds that can help anchor people in a virtual world and an audio source can sound like it comes from a specific object. As you build the soundscape for your app, consider using both types of audio.

Consider defining a range of places from which your app sounds can originate. Spatial Audio helps people locate the object that's making sound, whether it's stationary or moving in space. For example, when people move an app window that's playing audio, the sound continues to come directly from the window, wherever people move it.

Consider varying sounds that people could perceive as repetitive over time. For example, the system subtly varies the pitch and volume of the virtual keyboard's sounds, suggesting the different sounds a physical keyboard can make as people naturally vary the speed and forcefulness of their typing. An efficient way to achieve a pleasing variation in sound is to randomize a sound file's pitch and volume during playback, instead of creating different files.

Decide whether you need to play sound that's fixed to the wearer or tracked by the wearer. People perceive fixed sound as if it's pointed at them, regardless of the direction they look or the virtual objects they move. In contrast, people tend to perceive tracked sound as coming from a particular object, so moving the object closer or farther away changes what they hear. In general, you want to use tracked sound to enhance the realism of your experience, but there could be cases where fixed sound is a good choice.

7.3 (Near) Future of VR and Audio

By most measures, VR is still in its infancy. Despite its rich history going back to the mid-twentieth century, VR only began to hit a commercial stride in mid 2010s. This was primarily due to the availability of new display, tracking, and computational technologies that pushed VR over a threshold of user experience. Since then, there have been major developments in all of those technologies with rapid iterations in VR systems. There are no definitive statistics for VR usage worldwide as of this writing, but the Meta Quest 2 alone had seen over 10 million sales within its first year.[3] This is particularly impressive given the relatively high cost of VR systems and their growing pains pertaining to comfort and usability. That being said, in Chapter 1, we identified one of the biggest barriers to VR adoption as a lack of content, which itself is partly rooted in the lack of intuitive and reliable tools and workflows available to content creators.

7.3.1 *Workflow Improvements*

History has shown us time and again that media platforms become indispensable when they can harness the collective ingenuity of a broad range of creators. We witnessed a relatively recent example of this in the field of mobile computing. When Apple released the first version of the iPhone back in 2007, it introduced a new interaction paradigm with multi-touch screens. That year, Apple offered some compelling use cases with in-house applications that utilized touch input. However, it was the introduction of an App Store for third-party developers that blew open the platform for innovative applications that go far beyond the singular vision of a company. Independent developers quickly leveraged the unique features of this new mobile platform in a range of applications from productivity and games to stargazing and music production. This adoption was powered by the robust APIs, developer tools, and documentation that Apple provided to creators.

VR similarly introduces a new interaction paradigm with novel use cases. Besides its applications in entertainment, which has been the primary focus of this book, VR finds use in numerous fields including sports (Miah, Fenton, and Chadwick 2020), education (Rojas-Sánchez, Palos-Sánchez, and Folgado-Fernández 2023), engineering (Berni and Borgianni 2020), and healthcare (Mäkinen et al. 2022). Despite the rapid expansion of VR across numerous disciplines, the content offerings in today's popular VR marketplaces have arguably yet to reach a critical mass. The further the VR industry can tap into the imagination of creatives, the more quickly it will unearth compelling applications that will make VR an indispensable platform.

This brings us back to the need for accessible development and creativity support tools for VR. In Chapter 1, we identified some of the challenges facing VR designers today with a particular focus on audio design. In most film editing tools and game engines, the current VR and spatial audio implementations come across as stopgap solutions with missing or ill-conceived features. In the near future, we expect to see new creativity support tools and techniques that are purpose-built for the immersive medium. These tools will not only be created specifically for immersive media design but will also function in an immersive context. We are already seeing examples of this in 3D modeling with apps like ShapesXR and Gravity Sketch, which allow 3D designs to be executed in VR and overcome limitations of lower-dimensional input systems like mice, keyboards, and tablets. An example of this in the audio domain is DearVR's Spatial Connect that allows designers to visualize and control the spatial mix of a DAW session in VR. There are also numerous artistic and academic projects that focus on VR audio design within VR (Willette et al. 2020; Bell 2023; Dziwis 2023).

In addition to resolving the dimensional disparity between the design process and its product, these tools will also eliminate the discontinuity between the development and testing phases of a VR experience. In most modern creativity support tools, such as DAWs, film and photo editing tools, and game engines, the design actions are often reflected as immediate changes in the work observed within the same platform (e.g., desktop computer, tablet). In current VR workflows, the designer often finds themselves constantly switching between developing on a desktop system and testing in VR. This not only prolongs the work involved but also disrupts the action-perception feedback loop inherent to many modern creative processes.

Another major barrier to VR content creation, regardless of the design modality, is a lack of comprehensive and reliable documentation. The rapid evolution of VR systems today makes it challenging for hardware and software manufacturers to maintain up-to-date manuals, tutorials, and examples. This often prompts users to seek guidance on user forums for Unity and Unreal Engine, online communities such as Reddit and Stack Overflow, and other platforms like GitHub and ChatGPT. Furthermore, rapid iterations in technology can shorten the lifecycle of projects as one or more hardware and software platforms employed in a VR experience become incompatible due to an update. This disproportionately impacts novice developers and independent content creators, who may not have the resources to diagnose and fix bugs that stem from platform updates. This is an area where a community and industry-driven effort to standardize formats, frameworks, and workflows will help VR become a more sustainable platform for creative work.

7.3.2 *Real-to-Virtual Continuum*

Immersive media systems increasingly support mixed realities where virtual objects and environments are intermixed with real ones (Çamcı and Granzow 2019). The amalgamation of real and virtual worlds has distinct implications for audio. When a virtual object is placed into a real environment using an XR system, its scale, orientation, and position is adjusted based on planar-tracking of the real environment that the user inhabits. The virtual object is further blended in by having it cast shadows on real surfaces or, in some cases, be affected by the lighting conditions of the environment. A corollary of this for virtual sound elements would be to subject them to the acoustics of the real environment through room effects and occlusion. In fact, the Vision Pro does exactly this by estimating the reverberation in physical space by detecting its walls and surfaces and applying this reverb to the virtual sound sources in real time.

A growing number of headphone and wearable technology manufacturers today offer head tracking in their products. Alongside VR systems, these devices have substantially expanded consumer access to spatial audio. The ability to render acoustically augmented spaces using these technologies will have numerous implications for film, games, and music production, where sound sources will seamlessly blend into the soundscape of the environment that the user inhabits. Paired with noise-cancellation technology, this will give users detailed control over their everyday auditory environments.

Such work is broadly defined as Audio Augmented Reality (AAR). An early application of AAR is audio guides used in museums, where visitors can approach an artwork and hear details about it through headphones (Bederson 1995). However, later research argued for tighter integration between the real auditory environment that surrounds the listener and the virtual audio elements superimposed on it (Härmä et al. 2004). This is more closely aligned with more recent perspectives on AR, which emphasize the spatial and context-dependent integration of real and virtual objects and spaces. Modern applications of AAR range from navigation and location awareness to education and telepresence (Yang, Barde, and Billinghurst 2022). There are also innovative use cases in entertainment, where audiences are brought into shared participatory AAR spaces for gamified and theatrical experiences (Nagele et al. 2021).

7.3.3 Social Immersion

A majority of the VR applications available today cater to individual users. In contrast, most modern computational media facilitate multi-user experiences. We use our phones to interact with others through messaging, video calls, and social media. We use our computers for multiplayer gaming, video conferencing, and teamwork. Even when we are physically alone while using these devices, we are often connected to other people over the network. There are already compelling multi-user applications for VR. However, we expect social and collaborative use cases to become more prevalent as the consumer adoption of VR expands. The greater integration of our surroundings in our virtual experiences—a trend that we observe in recent VR systems—allows us to be increasingly engaged with our physical realities while in VR. This will support VR applications that rely on both in-person and remote social interactions.

The role of sound in social VR experiences is evident. Verbal interactions in VR not only convey meaning but also provide spatial cues and help users orient themselves to one another. The virtual acoustics of a social VR experience offer its users a shared context that extends beyond their field of view. But beyond sound, music is poised to be among the foremost applications of social VR. Musical creativity is often a collaborative activity that we engage in with or for other people. It prompts us to interact with others, our instruments, and our environments, while stimulating our senses of hearing, sight, touch, kinesthesia, and proprioception. Therefore, musical creativity has the capacity to push social VR to its limits. There are already numerous academic (Vilaplana Stark and Çamcı 2024) and commercial[4] applications of musical collaboration in VR. These bring together performers and audiences in real, virtual, and hybrid spaces. Moreover, VR can yield greater agency to the audience through participatory musical experiences (Çamcı and Hamilton 2020). These applications further underscore the role of sound design and, particularly, procedural audio in shared VR spaces.

7.3.4 An Intersection of Disciplines

One of the most exciting aspects of VR as a creative medium is its inherently interdisciplinary nature, where numerous areas of research and creative practice come together in their effort to move immersive technology forward. This intersection of disciplines is reminiscent of "the cognitive revolution" that emerged in the 1950s, when seemingly disparate fields like computer science, linguistics, philosophy, and psychology joined forces to further our understanding of the human mind (Miller 2003).

Throughout this book, we have highlighted the fusion of music, film, and games in VR. On a more technical level, perception studies are crucial for understanding our relationship with reality so that we can emulate it computationally in VR. Our engagement with this emulation draws heavily from work on human-computer interaction. And we are seeing an increasing use of artificial intelligence in VR—from the use of machine learning to detect users' hands as an input mechanism to the incorporation of large language models into responsive and generative VR environments. Other branches of computer science, such as simulation and networking, play a pivotal role in the design of complex, shared virtual environments. Industrial design focuses on the ergonomics of VR systems to ensure intuitive and comfortable user experiences from a hardware standpoint. Visual arts and architecture are intertwined in the design of rich virtual environments, involving models, structures, and landscapes with visual and spatial features that make VR experiences compelling. Lastly, the growing emphasis on ethics in immersive media studies highlights a range of critical issues, from privacy and accessibility to the potential psychological and developmental effects of VR.

It will be the collective effort of all these disciplines that ushers VR into its next generation. In that effort, we anticipate that many practitioners will broaden their skill sets to fully harness the unique offerings of VR, and we hope that this book supports that process from an audio perspective. Most importantly, we look to new creatives—such as those reading this book—to leverage the immersive medium to an extent we have yet to imagine.

Notes

1 https://developer.oculus.com/resources/bp-audio/. Accessed on July 19, 2024.
2 https://developer.apple.com/design/human-interface-guidelines/playing-audio#visionOS. Accessed on November 1, 2024.
3 https://www.theverge.com/2021/11/16/22785469/meta-oculus-quest-2-10-million-units-sold-qualcomm-xr2.
4 In PatchWorld (https://patchxr.com), multiple users can collaboratively perform with custom VR instruments. In VirtuosoVR (https://virtuoso-vr.com), a user can play various VR instruments while streaming their performance to an audience.

Bibliography

Abrash, Michael. 2014. "What VR Could, Should, and Almost Certainly Will Be Within Two Years." Steam Dev Days 2014, Seattle, Washington. https://www.youtube.com/watch?v=G-2dQoeqVVo.

Affel, H.A., R.W. Chesnut, and R.H. Mills. 1934. "Auditory Perspective—Transmission Lines." *Electrical Engineering*, 28–32, 214–16.

Agrawal, Sarvesh, Adèle Simon, Søren Bech, Klaus Bærentsen, and Søren Forchhammer. 2020. "Defining Immersion: Literature Review and Implications for Research on Audiovisual Experiences." *Journal of the Audio Engineering Society* 68 (6): 404–17.

Alais, David, and David Burr. 2004. "The Ventriloquist Effect Results from Near-Optimal Bimodal Integration." *Current Biology* 14 (3): 257–62.

Algazi, Ralph, Richard O. Duda, Dennis M. Thompson, and Carlos Avendano. 2001. "The CIPIC HRTF Database." In *Proceedings of the 2001 IEEE Workshop on the Applications of Signal Processing to Audio and Acoustics (Cat. No.01TH8575)*, 99–102. New Platz, NY, USA. doi:10.1109/ASPAA.2001.969552.

Algazi, V. Ralph, Richard O. Duda, and Dennis M. Thompson. 2004. "Motion-Tracked Binaural Sound." *Journal of the Audio Engineering Society* 52 (11): 1142–56.

Allen, Jont B, and David A. Berkley. 1979. "Image Method for Efficiently Simulating Small-Room Acoustics." *Journal of the Acoustical Society of America* 65 (4): 943–50.

Allen, Jont B. 1996. "Harvey Fletcher's Role in the Creation of Communication Acoustics." *The Journal of the Acoustical Society of America* 99 (4): 1825–39.

Ament, Vanessa Theme. 2014. *The Foley Grail: The Art of Performing Sound for Film, Games, and Animation*. Routledge.

"Augmented and Virtual Reality Survey Report." 2020. 4. Industry Insights into the Future of Immersive Technology. Perkins Coie LLP.

Bailey, Will, and Bruno M. Fazenda. 2017. "The Effect of Reverberation and Audio Spatialization on Egocentric Distance Estimation of Objects in Stereoscopic Virtual Reality." *Journal of the Acoustical Society of America* 141 (5_Supplement) (May): 3510.

Barnwell, Jane. 2019. *The Fundamentals of Film Making*. Bloomsbury Publishing.

Barrett, Natasha. 2002. "Spatio-Musical Composition Strategies." *Organised Sound* 7 (3): 313–23.

Barron, M., and A.H. Marshall. 1981. "Spatial Impression Due to Early Lateral Reflections in Concert Halls: The Derivation of a Physical Measure." *Journal of Sound and Vibration* 77 (2): 211–32. https://doi.org/10.1016/S0022-460X(81)80020-X.

Bates, Enda. 2018. "A Recording Technique for 6 Degrees of Freedom VR." Paper presented at *144th Convention of the Audio Engineering Society*, Milan, Italy. *144th Convention of the Audio Engineering Society*.

Beauchemin, Stéphane. 2016. "Wwise and the Audio Programmer." In *Game Audio Programming*, edited by Guy Somberg. CRC Press.

Bédard, Philippe. 2023. "Closing the Gap: Storytelling in Nonfiction Virtual Reality." *Canadian Journal of Film Studies/Revue Canadienne D′ Études Cinématographiques* 32 (1): 87–105.

Bederson, Benjamin B. 1995. "Audio Augmented Reality: A Prototype Automated Tour Guide." In *Conference Companion on Human Factors in Computing Systems*, 210–211.

Begault, Durand R. 2000. "3-D Sound for Virtual Reality and Multimedia." NASA Ames Research Center.

Bell, Jonathan. 2023. "Networked Music Performance in PatchXR and FluCoMa." In *Proceedings of the International Computer Music Conference (ICMC) 2023*.

Bellinger, Daniel, Kristin Wehrmann, Anna Rohde, Maria Schuppert, Stefan Störk, Michael Flohr-Jost, Dominik Gall, Paul Pauli, Jürgen Deckert, and Martin J. Herrmann. 2023. "The Application of Virtual Reality Exposure Versus Relaxation Training in Music Performance Anxiety: A Randomized Controlled Study." *BMC Psychiatry* 23 (555).

Belton, John. 1985. "Technology and Aesthetics of Film Sound." In *Film Sound: Theory and Practice*, edited by Elisabeth Weis and John Belton. Columbia University Press.

Benjamin, Eric. 2012. "A Second-Order Soundfield Microphone with Improved Polar Pattern Shape." In *Proceedings of the 133rd Audio Engineering Society Convention*. Audio Engineering Society.

Berkhout, Augustinus J. 1988. "A Holographic Approach to Acoustic Control." *Journal of the Audio Engineering Society* 36 (12): 977–95.

Berni, Aurora, and Yuri Borgianni. 2020. "Applications of Virtual Reality in Engineering and Product Design: Why, What, How, When and Where." *Electronics* 9 (7): 1064.

Bertet, Stéphanie, Jérôme Daniel, Etienne Parizet, and Olivier Warusfel. 2013. "Investigation on Localisation Accuracy for First and Higher Order Ambisonics Reproduced Sound Sources." *Acta Acustica United with Acustica* 99 (4): 642–57.

Best, Virginia, Robert Baumgartner, Mathieu Lavandier, Piotr Majdak, and Norbert Kopco. 2020. "Sound Externalization: A Review of Recent Research." *Trends in Hearing* 24. https://doi.org/10.1177/2331216520948390.

Bissonnette, Josiane, Francis Dubé, Martin D. Provencher, and Maria T. Moreno Sala. 2016. "Evolution of Music Performance Anxiety and Quality of Performance During Virtual Reality Exposure Training." *Virtual Reality* 20:71–81.

Blumlein, Alan Dower. 1958. "British Patent Specification 394,325." *Journal of the Audio Engineering Society* 6 (2): 91–98.

Boone, Andrew R. 1941. "Mickey Mouse Goes Classical." *Popular Science*, January 1941. https://books.google.com/books?id=NicDAAAAMBAJ&pg=PA65#v=onepage&q&f=false.

Bossalini, Cameron, William Raffe, and Jaime Andres Garcia. 2020. "Generative Audio and Real-Time Soundtrack Synthesis in Gaming Environments: An Exploration of How Dynamically Rendered Soundtracks Can Introduce New Artistic Sound Design Opportunities and Enhance the Immersion of Interactive Audio Spaces." In *OzCHI '20: Proceedings of the 32nd Australian Conference on Human-Computer Interaction*, 281–292.

Boyer, Steven. 2009. "A Virtual Failure: Evaluating the Success of Nintendo's Virtual Boy." *The Velvet Light Trap*, no. 64, 23–33.

Brandenburg, Karlheinz, Carmelo Fascella, Nils Merten, et al. 2023. "Implementation of and Application Scenarios for Plausible Immersive Audio via Headphones." *Presented at the 155th Convention of the Audio Engineering Society* (New York, NY).

Brey, Philip. 2020. "The Ethics of Representation and Action in Virtual Reality." In *The Ethics of Information Technologies*, 405–14. Routledge.

Brice, Richard. 2019. "A New Decoder for CD-4 (Quadradisc) Phonograph Records." *Journal of the Audio Engineering Society* 67 (9): 679–90. https://doi.org/10.17743/jaes.2019.0016.

Brimijoin, II, William O., Henrik G. Hassager, Vamsi K. Ithapu, and Philip Robinson. 2020. Individualization of head related transfer function templates for presentation of audio content. United States Patent Office US 2020/0336858 A1, issued October 22, 2020.

Broderick, James, Jim Duggan, and Sam Redfern. 2018. "The Importance of Spatial Audio in Modern Games and Virtual Environments." In *Proceedings of the 2018 IEEE Games, Entertainment, Media (GEM) Conference*, 1–9. IEEE.

Bruin, S.L. de. 1958. "The 'Electronic Poem' Performed in the Philips Pavilion at the 1958 Brussels World Fair: C. The Electronic Control System." *Philips Technical Review* 20 (2/3): 45–49.

Brumm, Adam, Adhi Agus Oktaviana, Basran Burhan, Budianto Hakim, Rustan Lebe, Jian-xin Zhao, Priyatno Hadi Sulistyarto, Marlon Ririmasse, Shinatria Adhityatama, and Iwan Sumantri. 2021. "Oldest Cave Art Found in Sulawesi." *Science Advances* 7 (3): eabd4648.

Brungart, Douglas S. 2002. "Near-Field Virtual Audio Displays." *Presence II* (I): 93–106.

Bujić, Mila, Mikko Salminen, Joseph Macey, and Juho Hamari. 2020. "'Empathy Machine': How Virtual Reality Affects Human Rights Attitudes." *Internet Research* 30 (5): 1407–25.

Butterworth, Alicia. 2022. "Beyond Sonic Realism: A Cinematic Sound Approach in Documentary 360 Film." *Studies in Documentary Film* 16 (2): 156–73.

Çamcı, Anıl, and John Granzow. 2019. "Hyperreal Instruments: Bridging VR and Digital Fabrication to Facilitate New Forms of Musical Expression." *Leonardo Music Journal* 29: 14–18.

Çamcı, Anıl, and Rob Hamilton. 2020. "Audio-First VR: New Perspectives on Musical Experiences in Virtual Environments." *Journal of New Music Research* 49 (1): 1–7.

Çamcı, Anıl, Kristine Lee, Cody J Roberts, and Angus G Forbes. 2017. "Inviso: A Cross-Platform User Interface for Creating Virtual Sonic Environments." In *Proceedings of the 30th Annual ACM Symposium on User Interface Software and Technology*, 507–518.

Çamcı, Anıl. 2019. "Exploring the Effects of Diegetic and Non-Diegetic Audiovisual Cues on Decision-Making in Virtual Reality." In *Proceedings of the 16th Sound and Music Computing Conference*, 28–31.

Çamcı, Anıl. 2022. *The Cognitive Continuum of Electronic Music*. Bloomsbury Publishing.

Çamcı, Anıl. 2024. "Modern Workflows for Procedural Audio at the Intersection of Gaming and Music Performance in Virtual Reality." In *Proceedings of the 6th AES International Conference on Audio for Games*. Tokyo, Japan: Audio Engineering Society.

Carpentier, Thibaut, Markus Noisternig, and Olivier Warusfel. 2015. "Twenty Years of IRCAM Spat: Looking Back, Looking Forward." In *41st International Computer Music Conference (ICMC)*, 270–277.

Collins, Karen. 2008. *Game Sound: An Introduction to the History, Theory, and Practice of Video Game Music and Sound Design*. Mit Press.

Collins, Karen. 2009. "An Introduction to Procedural Music in Video Games." *Contemporary Music Review* 28 (1): 5–15.

Corey, Jason, and Wieslaw Woszczyk. 2002. "Localization of Lateral Phantom Images in a 5-Channel System with and without Simulated Early Reflections." Los Angeles, CA: Presented at the *113th Convention of the Audio Engineering Society*.

Corey, Jason, Wieslaw Woszczyk, René Quesnel, and Geoff Martin. 2001. "Enhancements of Room Simulation with Dynamic Cues Related to Source Position." In *Proceedings of the Audio Engineering Society 19th International Conference: Surround Sound - Techniques, Technology, and Perception*. Schloss Elmau, Germany: Audio Engineering Society.

Cruz-Neira, Carolina, Daniel J Sandin, and Thomas A DeFanti. 1993. "Surround-Screen Projection-Based Virtual Reality: The Design and Implementation of the CAVE." In *Proceedings of the 20th Annual Conference on Computer Graphics and Interactive Techniques*, 135–142.

Dienstfrey, Eric. 2016. "The Myth of the Speakers: A Critical Reexamination of Dolby History." *Film History* 28 (1): 167–193. https://doi.org/10.2979/filmhistory.28.1.06.

"Dolby Digital 5.1." n.d. Corporate. Dolby Digital 5.1. Accessed August 1, 2023. https://professional.dolby.com/tv/dolby-digital/.

Dolby Laboratories, Inc. 2022. "Dolby Atmos Personalized Rendering (Beta)." *Dolby Professional (blog)*. 2022. https://professional.dolby.com/phrtf/.

Dolby Laboratories, Inc. n.d. "7.1.4 Overhead Speaker Setup." Dolby Atmos Speaker Setup Guides. https://www.dolby.com/about/support/guide/speaker-setup-guides/7.1.4-overhead-speaker-setup-guide.

Dooley, Kath. 2021. *Cinematic Virtual Reality*. Springer.
Dziwis, Damian. 2023. "PdXR-the Evolution of Pure Data into the Metaverse." In *Proceedings of the International Computer Music Conference (ICMC) 2023*, 1–8.
Eargle, John. 1976. *Sound Recording*. New York, NY: Van Nostrand Reinhold Company.
Ekman, Inger. 2005. "Meaningful Noise: Understanding Sound Effects in Computer Games." *Proceedings of Digital Arts and Cultures* 17.
Faller, Christof, Lutz Altmann, Jeff Levison, and Markus Schmidt. 2013. "Multi-Channel Ring Upmix." Rome, Italy: Presented at the *134th Convention of the Audio Engineering Society, Convention Paper 8908*.
Farnell, Andy. 2014. "Procedural Audio Theory and Practice." In *The Oxford Handbook of Interactive Audio*, edited by Kasey Collins, Bill Kapralos, and Holly Tessler. Oxford University Press. https://doi.org/10.1093/oxfordhb/9780199797226.013.031.
Feltham, Jamie. 2021. "Oculus Quest 2 Is Now the Most-Used VR Headset on Steam." *UploadVR (blog)*. March 2, 2021. https://www.uploadvr.com/oculus-quest-2-steam-most-used/.
Fichna, Stefan, Steven van de Par, and Stephan D. Ewert. 2023. "Evaluation of Virtual Acoustic Environments with Different Acoustic Level of Detail." In *2023 Immersive and 3D Audio: From Architecture to Automotive (I3DA)*. Bologna, Italy. https://doi.org/10.1109/I3DA57090.2023.10289496.
Fine, Robert. 1954. "PERSPECTA—the All-Purpose Recording and Reproducing Sound System." *International Projectionist* 29 (7): 32–33, 41–42.
Fischer, Tim, Marco Caversaccio, and Wilhelm Wimmer. 2020. "A Front-Back Confusion Metric in Horizontal Sound Localization: The FBC Score." In *ACM Symposium on Applied Perception 2020 (SAP '20)*, Article 11:1–5. New York, NY, USA: Association for Computing Machinery. https://doi.org/10.1145/3385955.3407928.
Fletcher, Harvey. 1934. "Auditory Perspective—Basic Requirements." *Electrical Engineering*, 9–11.
Fluitt, Kim, Tomasz Letowski, and Timothy Mermagen. 2014. "Auditory Distance Estimation in an Open Space." In *Soundscape Semiotics–Localization and Categorization*, edited by Hervé Glotin. Rijeka: IntechOpen. https://doi.org/10.5772/56137.
Gallacher, Nicola. 2013. "Game Audio—An Investigation into the Effect of Audio on Player Immersion." *The Computer Games Journal* 2:52–79.
Gardner, William G., and Keith D. Martin. 1995. "HRTF Measurements of a KEMAR." *Journal of the Acoustical Society of America* 97 (6): 3907–8.
Garity, W. E., and J. N. Hawkins. 1941. "Fantasound." *Journal of the Society of Motion Picture Engineers* 37 (8): 127–46.
Gazdag, Jeno. 1981. "Modeling of the Acoustic Wave Equation with Transform Methods." *Geophysics* 46 (6): 854–59. https://doi.org/10.1190/1.1441223.
Gear, Gavin. 2013. "Hands-On with Oculus Rift Virtual Reality Development Kit." *Windows Blogs (blog)*. December 11, 2013. https://blogs.windows.com/windowsexperience/2013/12/11/hands-on-with-oculus-rift-virtual-reality-development-kit/.
Genette, Gérard. 1983. *Narrative Discourse: An Essay in Method*. Vol. 3. Cornell University Press.
Genzel, Daria, Michael Schutte, W. Owen Brimijoin, Paul R. MacNeilage, and Lutz Wiegrebe. 2018. "Psychophysical Evidence for Auditory Motion Parallax." *Proceedings of the National Academy of Sciences* 115 (16): 4264–69.
Geronazzo, Michael, Amalie Rosenkvist, David Sebastian Eriksen, et al. 2019. "Creating an Audio Story with Interactive Binaural Rendering in Virtual Reality." *Wireless Communications and Mobile Computing* 2019 (Article ID 1463204): 14 pages. https://doi.org/10.1155/2019/1463204.
Gerzon, Michael. 1971. "Experimental Tetrahedral Recording: Part One." *Studio Sound*, August 1971. Retrieved from Into The Soundfield: Michael Gerzon and Ambisonics at Oxford: https://intothesoundfield.music.ox.ac.uk/sites/default/files/intothesoundfield/documents/media/tetrahedral_1.pdf.

Gerzon, Michael. 1985. "Ambisonics in Multichannel Broadcasting and Video." *Journal of the Audio Engineering Society* 33 (11): 859–71.

Gilkey, Robert H., and Janet M. Weisenberger. 1995. "The Sense of Presence for the Suddenly Deafened Adult: Implications for Virtual Environments." *Presence: Teleoperators and Virtual Environments* 4 (4): 357–63.

Görne, Thomas. 2019. "The Emotional Impact of Sound: A Short Theory of Film Sound Design." *EPiC Series in Technology* 1:17–30.

Gorzel, Marcin, Andrew Allen, Ian Kelly, Julius Kammerl, Alper Gungormusler, Hengchin Yeh, and Francis Boland. 2019. "Efficient Encoding and Decoding of Binaural Sound with Resonance Audio." In *Proceedings of the 2019 Aes International Conference on Immersive and Interactive Audio*. Audio Engineering Society.

Hacihabiboglu, Huseyin, Enzo De Sena, Zoran Cvetkovic, James Johnston, and Julius O. Smith III. 2017. "Perceptual Spatial Audio Recording, Simulation, and Rendering: An Overview of Spatial-Audio Techniques Based on Psychoacoustics." *IEEE Signal Processing Magazine* 34 (3): 36–54. https://doi.org/10.1109/MSP.2017.2666081.

Hadan, Hilda, Lydia Choong, Leah Zhang-Kennedy, and Lennart E Nacke. 2024. "Deceived by Immersion: A Systematic Analysis of Deceptive Design in Extended Reality." *ACM Computing Surveys* 56 (10): 1–25.

Hamasaki, Kimio, and Wilfried Van Baelen. 2015. "Natural Sound Recording of an Orchestra with Three-Dimensional Sound." *138th Convention of the Audio Engineering Society Convention Paper 9348*.

Hamasaki, Kimio, Koichiro Hiyama, Toshiyuki Nishiguchi, and Kazuho Ono. 2004. "Advanced Multichannel Audio Systems with Superior Impression of Presence and Reality." Berlin, Germany: Presented at the *116th Convention of the Audio Engineering Society, Convention Paper 6053*.

Hamasaki, Kimio, Toshiyuki Nishiguchi, Reiko Okumura, Yasushige Nakayama, and Akio Ando. 2008. "A 22.2 Multichannel Sound System for Ultrahigh-Definition TV (UHDTV)." *SMPTE Motion Imaging Journal* 117 (3): 40–49.

Härmä, Aki, Julia Jakka, Miikka Tikander, Matti Karjalainen, Tapio Lokki, Jarmo Hiipakka, and Gaëtan Lorho. 2004. "Augmented Reality Audio for Mobile and Wearable Appliances." *Journal of the Audio Engineering Society* 52 (6): 618–39.

Hartmann, William M. 2021. "Localization and Lateralization of Sound." In *Binaural Hearing. Springer Handbook of Auditory Research*, edited by R.Y. Litovsky, M.J. Goupell, R.R. Fay, and A.N. Popper. Vol. 73. Springer, Cham. https://doi.org/10.1007/978-3-030-57100-9_2.

Heilig, Morton L. 1960. "Stereoscopic-Television Apparatus for Individual Use," US Patent 2955156A, issued 4 October 1960.

Herre, Jürgen, and Sascha Disch. 2023. "MPEG-I Immersive Audio - Reference Model for the Virtual/Augmented Reality Audio Standard." *Journal of the Audio Engineering Society* 71 (5): 229–40. https://doi.org/10.17743/jaes.2022.0074.

Herre, Jürgen, Johannes Hilpert, Achim Kuntz, and Jan Plogsties. 2014. "MPEG-H Audio—The New Standard for Universal Spatial/3D Audio Coding." *Journal of the Audio Engineering Society* 62 (12): 821–30.

Herre, Jürgen, Sascha Disch, Christian Borss, Andreas Silzle, Alexander Adami, and Nils Peters. 2024. "MPEG-I Immersive Audio: A Versatile and Efficient Representation of VR/AR Audio Beyond Point Source Rendering." Paper presented at Audio Engineering Society International Conference on Audio for Games, Tokyo, Japan. *Audio Engineering Society International Conference on Audio for Games*, April 27.

Hodsdon, Barrett. 1992. "The Mystique of Mise En Scene Revisited." *Continuum: Journal of Media & Cultural Studies* 5 (2): 68–86.

Holman, Tomlinson. 2000. *5.1 Surround Sound: Up and Running*. Burlington, MA: Focal Press.

Hong, Joo Young, Bhan Lam, Zhen-Ting Ong, et al. 2019. "Quality Assessment of Acoustic Environment Reproduction Methods for Cinematic Virtual Reality in Soundscape Applications." *Building and Environment* 149: 1–14. https://doi.org/10.1016/j.buildenv.2018.12.004.

Horowitz, Steve, and Scott Looney. 2014. *The Essential Guide to Game Audio: The Theory and Practice of Sound for Games*. Routledge.

Huang, Yi-Hsuan, Roshan Venkatakrishnan, Rohith Venkatakrishnan, Sabarish V Babu, and Wen-Chieh Lin. 2021. "Using Audio Reverberation to Compensate Distance Compression in Virtual Reality." *SAP '21: ACM Symposium on Applied Perception 2021* (Virtual Event, France). https://doi.org/10.1145/3474451.3476236.

IJsselsteijn, Wijnand A., Huib de Ridder, Jonathan Freeman, and Steve E. Avons. 2000. "Presence: Concept, Determinants and Measurement." *Proceedings of SPIE 3959, Human Vision and Electronic Imaging V* (San Jose, CA), 520–529. https://doi.org/10.1117/12.387188.

International Telecommunication Union. 2019. *Recommendation ITU-R BS.2076-2: Audio Definition Model*. ITU. https://www.itu.int/rec/R-REC-BS.2076.

International Telecommunication Union. 2022. "Recommendation ITU-R BS.775-4: Multichannel Stereophonic Sound System with and without Accompanying Picture." Geneva, Switzerland: ITU.

International Telecommunication Union. 2023. *Recommendation ITU-R BS.2127-1: Audio Definition Model Renderer for Advanced Sound Systems*. ITU. https://www.itu.int/rec/R-REC-BS.2127.

Islam, Md Tamzeed, and Ivan J. Tashev. 2020. "Anthropometric Features Estimation Using Integrated Sensors on a Headphone for HRTF Personalization." In *2020 AES International Conference on Audio for Virtual and Augmented Reality*. Redmond, Washington: Audio Engineering Society.

ITU-R. 2012. "Multichannel Sound Technology in Home and Broadcasting Applications." *ITU-R BS.2159-4. Broadcasting Service*. Geneva, Switzerland: International Telecommunication Union.

Jenny, Claudia, and Christoph Reuter. 2020. "Usability of Individualized Head-Related Transfer Functions in Virtual Reality: Empirical Study with Perceptual Attributes in Sagittal Plane Sound Localization." *JMIR Serious Games* 8 (3).

Jones, Brett R, Hrvoje Benko, Eyal Ofek, and Andrew D Wilson. 2013. "IllumiRoom: Peripheral Projected Illusions for Interactive Experiences." In *Proceedings of the SIGCHI Conference on Human Factors in Computing Systems*, 869–78.

Jot, Jean-Marc, René Audfray, Mark Hertensteiner, and Brian Schmidt. 2021. "Rendering Spatial Sound for Interoperable Experiences in the Audio Metaverse." In *Proceedings of the International Conference on Immersive and 3D Audio: From Architecture to Automotive (I3DA)*. Virtual (Bologna, Italy): IEEE.

Kailas, Ganesh, and Nachiketa Tiwari. 2021. "Design for Immersive Experience: Role of Spatial Audio in Extended Reality Applications." In *Design for Tomorrow—Volume 2. Smart Innovation, Systems and Technologies*, edited by A. Chakrabarti, R. Poovaiah, P. Bokil, and V. Kant, vol. 222. Springer, Singapore. https://doi.org/10.1007/978-981-16-0119-4_69.

Kaimara, Polyxeni, Andreas Oikonomou, and Ioannis Deliyannis. 2022. "Could Virtual Reality Applications Pose Real Risks to Children and Adolescents? A Systematic Review of Ethical Issues and Concerns." *Virtual Reality* 26 (2): 697–735.

Kastbauer, Damian. 2014. *The Wwise Project Adventure: A Handbook for Creating Interactive Audio Using Wwise*. Audiokinetic. https://www.audiokinetic.com/download/documents/WwiseProjectAdventure_en.pdf.

Kelly, Jack, Wieslaw Woszczyk, and Richard King. 2020. "Are You There?: A Literature Review of Presence for Immersive Music Reproduction." *149th Convention of the Audio Engineering Society (Online)*, Convention Paper 10389.

Klapholz, Jesse. 1991. "Fantasia: Innovations in Sound." *Journal of the Audio Engineering Society* 39 (1/2): 66–70.

Klein, Florian, Stephan Werner, Georg Götz, and Karlheinz Brandenburg. 2019. "Auditory Adaptation in Real and Virtual Rooms." In *International Symposium on Auditory and Audiological Research (Proc. ISAAR), Vol. 7: Auditory Learning in Biological and Artificial Systems*, 342–48. Nyborg, Denmark: The Danavox Jubilee Foundation.

Krueger, Myron W. 1977. "Responsive Environments." In *Proceedings of the National Computer Conference*, 423–33.

Lanier, Jaron. 1992. "Virtual Reality: The Promise of the Future." *Interactive Learning International* 8 (4): 275–79.

Larsson, Pontus, Aleksander Väljamäe, Daniel Västfjäll, Ana Tajadura-Jiménez, and Mendel Kleiner. 2010. "Auditory-Induced Presence in Mixed Reality Environments and Related Technology." In *The Engineering of Mixed Reality Systems*, edited by Emmanuel Dubois, Philip Gray, and Laurence Nigay. Human-Computer Interaction Series. Springer-Verlag.

Lee, Hyunkook. 2019. "Capturing 360° Audio Using an Equal Segment Microphone Array (ESMA)." *Journal of the Audio Engineering Society* 67 (1/2): 13–26.

Lee, Hyunkook. 2021. "Multichannel 3D Microphone Arrays: A Review." *Journal of the Audio Engineering Society* 69 (1/2): 5–26.

Lindquist, Mark, Bruce Maxim, Jennifer Proctor, and Francine Dolins. 2020. "The Effect of Audio Fidelity and Virtual Reality on the Perception of Virtual Greenspace." *Landscape and Urban Planning* 202: 103884. https://doi.org/10.1016/j.landurbplan.2020.103884.

Lockyer, Norman. 1881. "The International Exhibition and Congress of Electricity at Paris, IV." *Nature* 24 (626): 585–89.

Lombard, Matthew, and Theresa Ditton. 1997. "At the Heart of It All: The Concept of Presence." *Journal of Computer-Mediated Communication* 3 (2): JCMC321. https://doi.org/10.1111/j.1083-6101.1997.tb00072.x.

Lopez, Oscar. 2011. "AD Classics: Expo '58 + Philips Pavilion / Le Corbusier and Iannis Xenakis." *ArchDaily*. August 25, 2011. https://www.archdaily.com/157658/ad-classics-expo-58-philips-pavilion-le-corbusier-and-iannis-xenakis.

Mäkinen, Henna, Elina Haavisto, Sara Havola, and Jaana-Maija Koivisto. 2022. "User Experiences of Virtual Reality Technologies for Healthcare in Learning: An Integrative Review." *Behaviour & Information Technology* 41 (1): 1–17.

Marks, Aaron. 2017. *Aaron Marks' Complete Guide to Game Audio: For Composers, Sound Designers, Musicians, and Game Developers*. CRC Press.

Marrinan, Thomas, Pakeeza Akram, Oli Gurmessa, and Anthony Shishkin. 2024. "Leveraging AI to Generate Audio for User-Generated Content in Video Games." *arXiv Preprint arXiv: 2404.17018*.

Martin, Geoff, Wieslaw Woszczyk, Jason Corey, and René Quesnel. 1999a. "Sound Source Localization in a Five-Channel Surround Sound Reproduction System." New York, NY: Presented at the *107th Convention of the Audio Engineering Society*.

Martin, Geoff, Wieslaw Woszczyk, Jason Corey, and René Quesnel. 1999b. "Sound Source Localization in a Five-Channel Surround Sound Reproduction System." *107th Convention of the Audio Engineering Society*, Convention Paper 4994.

McArthur, Angela, Rebecca Stewart, and Mark Sandler. 2017. "Sounds Too True to Be Good: Diegetic Infidelity–the Case for Sound in Virtual Reality." *Journal of Media Practice* 18 (1): 26–40.

McCormack, Leo, Archontis Politis, Thomas McKenzie, Christoph Hold, and Ville Pulkki. 2022. "Object-Based Six-Degrees-of-Freedom Rendering of Sound Scenes Captured with Multiple Ambisonic Receivers." *Journal of the Audio Engineering Society* 70 (5): 355–72. https://doi.org/10.17743/jaes.2022.0010.

Melchior, Frank. 2010. "Wave Field Synthesis and Object-Based Mixing for Motion Picture Sound." *SMPTE Motion Imaging Journal*, no. April, 53–57.

Meta. 2023. "Improve Spatial Audio Quality with the Universal HRTF." *Meta Quest News (blog)*. July 11, 2023. https://developer.oculus.com/blog/improve-spatial-audio-universal-hrtf-meta-quest/.

Meta. 2024. "Immersive, Realistic Audio in VR: Introducing Acoustic Ray Tracing in Audio SDK." *Blog. Meta Quest News (blog)*. July 30, 2024. https://developer.oculus.com/blog/acoustic-ray-tracing-audio-sdk-meta-quest-developer-social-presence/.

Meyer, Jürgen. 2009a. *Acoustics and the Performance of Music: Manual for Acousticians, Audio Engineers, Musicians, Architects and Musical Instrument Makers*. Translated by Uwe Hansen. Fifth Edition. New York, NY: Springer.

Meyer, Jürgen. 2009b. *Acoustics and the Performance of Music: Manual for Acousticians, Audio Engineers, Musicians, Architects and Musical Instrument Makers*. Fifth Edition. Translated by Uwe Hansen. Springer.

Miah, Andy, Alex Fenton, and Simon Chadwick. 2020. "Virtual Reality and Sports: The Rise of Mixed, Augmented, Immersive, and Esports Experiences." *21st Century Sports: How Technologies Will Change Sports in the Digital Age*, 249–62.

Milgram, Paul, and Fumio Kishino. 1994. "A Taxonomy of Mixed Reality Visual Displays." *IEICE Transactions on Information and Systems* 77 (12): 1321–29.

Miller, George A. 2003. "The Cognitive Revolution: A Historical Perspective." *Trends in Cognitive Sciences* 7 (3): 141–44.

Morimoto, Masayuki, and Yoichi Ando. 1980. "On the Simulation of Sound Localization." *Journal of the Acoustical Society of Japan* 1 (3): 167–74.

Mróz, B., M. Kabaciński, T. Ciotucha, A. Rumiński, and T. Żernicki. 2021. "Production of Six-Degrees-of-Freedom (6DoF) Navigable Audio Using 30 Ambisonic Microphones." *2021 Immersive and 3D Audio: From Architecture to Automotive (I3DA)*, September 8, 1–5. https://doi.org/10.1109/I3DA48870.2021.9610970.

Murch, Walter. 2005. "Part 2: Dense Clarity - Clear Density." *Transom (blog)* 2005. https://transom.org/2005/walter-murch/.

Murray, Jeff W. 2021. *C# Game Programming Cookbook for Unity 3d*. CRC Press.

Mütterlein, Joschka. 2018. "The Three Pillars of Virtual Reality?: Investigating the Roles of Immersion, Presence, and Interactivity." *Proceedings of the 51st Hawaii International Conference on System Sciences*. http://hdl.handle.net/10125/50061.

Nagele, Anna N., Valentin Bauer, Patrick G.T. Healey, Joshua D. Reiss, Henry Cooke, Tim Cowlishaw, Chris Baume, and Chris Pike. 2021. "Interactive Audio Augmented Reality in Participatory Performance." *Frontiers in Virtual Reality* 1:610320.

Neidhardt, Annika. 2023. "On the Plausibility of Simplified Acoustic Room Representations for Listener Translation in Dynamic Binaural Auralizations." Doctoral Dissertation, Ilmenau: Technische Universität Ilmenau. https://doi.org/10.22032/dbt.57596.

Nicol, Rozenn. 2020. "Creating Auditory Illusions with Spatial-Audio Technologies." In *The Technology of Binaural Understanding*, edited by Jens Blauert and Jonas Braasch. Modern Acoustics and Signal Processing. Springer Nature Switzerland AG.

Oculus VR. 2017. "Beyond Surround Sound: Audio Advances in VR." *Meta Quest Blog (blog)*. September 19, 2017. https://www.meta.com/blog/quest/beyond-surround-sound-audio-advances-in-vr/.

Olivieri, Ferdinando, Nils Peters, and Deep Sen. 2019. *Scene-Based Audio and Higher Order Ambisonics: A Technology Overview and Application to Next-Generation Audio, VR and 360° Video*. Technical Review. European Broadcasting Union.

Park, Jieun, Youjin Choi, and Kyung Myun Lee. 2024. "Research Trends in Virtual Reality Music Concert Technology: A Systematic Literature Review." *IEEE Transactions on Visualization and Computer Graphics*, 2195–2205.

Potter, Thomas, Zoran Cvetkovic, and Enzo De Sena. 2022. "On the Relative Importance of Visual and Spatial Audio Rendering on VR Immersion." *Frontiers in Signal Processing* 2: 904866. https://doi.org/10.3389/frsip.2022.904866.

Pulkki, Ville. 2001. "Spatial Sound Generation and Perception by Amplitude Panning Techniques." PhD Dissertation, Helsinki University of Technology.

Ramsdell, D.A. 1970. "The Psychology of the Hard-of-Hearing and the Deafened Adult." In *Hearing and Deafness*, by H. Davis and S.R. Silverman, 3rd ed. New York, NY: Rinehart and Winston.

Rayleigh, L. 1907. "XII. On Our Perception of Sound Direction." *The London, Edinburgh, and Dublin Philosophical Magazine and Journal of Science* 13 (74): 214–32. https://doi.org/10.1080/14786440709463595.

Reardon, Gregory, Andrea Genovese, Gabriel Zalles, Patrick Flanagan, and Agnieszka Roginska. 2018a. "Evaluation of Binaural Renderers: Localization." *144th Convention of the Audio Engineering Society Convention Paper 9963*.

Reardon, Gregory, Andrea Genovese, Gabriel Zalles, Patrick Flanagan, and Agnieszka Roginska. 2018b. "Evaluation of Binaural Renderers: Multidimensional Quality Assessment." Paper presented at Audio Engineering Society International Conference on Audio for Virtual and Augmented Reality, Redmond, Washington. *Audio Engineering Society International Conference on Audio for Virtual and Augmented Reality*.

Reardon, Gregory, Gabriel Zalles, Andrea Genovese, Patrick Flanagan, and Agnieszka Roginska. 2018. "Evaluation of Binaural Renderers: Externalization, Front/Back and Up/Down Confusions." Paper presented at *144th Convention of the Audio Engineering Society*, Milan, Italy. *144th Convention of the Audio Engineering Society*.

Robinson, Ciarán. 2019. *Game Audio with FMOD and Unity*. Routledge.

Rojas-Sánchez, Mario A., Pedro R Palos-Sánchez, and José A. Folgado-Fernández. 2023. "Systematic Literature Review and Bibliometric Analysis on Virtual Reality and Education." *Education and Information Technologies* 28 (1): 155–92.

Rummukainen, Olli S., Thomas Robotham, and Emanuël A. P. Habets. 2021. "Head-Related Transfer Functions for Dynamic Listeners in Virtual Reality." *Applied Sciences* 11 (14): 6646. https://doi.org/10.3390/app11146646.

Satongar, Darius A., Jonathan D. Sheaffer, Martin E. Johnson, and Peter V. Jupin. 2020. Head related transfer function selection for binaural sound reproduction. United States Patent Office US10728690B1, issued July 28, 2020.

Savioja, Lauri, and U. Peter Svensson. 2015. "Overview of Geometrical Room Acoustic Modeling Techniques." *The Journal of the Acoustical Society of America* 138 (2): 708–30. https://doi.org/10.1121/1.4926438.

Scheiber, Peter. 2003. "Comments on 'History of Spatial Coding." *Journal of the Audio Engineering Society* 51 (11): 1062.

Schutte, Nicola S., and Emma J. Stilinović. 2017. "Facilitating Empathy Through Virtual Reality." *Motivation and Emotion* 41:708–12.

Scientific American. 1881. "The Telephone at the Paris Opera," December 31, 1881.

Shim, Hwan, Eunmi Oh, Sangchul Ko, and Sang Ha Park. 2010. "Perceptual Evaluation of Spatial Audio Quality." San Francisco, California: Presented at the *129th Convention of the Audio Engineering Society, Convention Paper 8300*.

Shivappa, Shankar, Martin Morrell, Deep Sen, Nils Peters, and S.M. Akramus Salehin. 2016. "Efficient, Compelling, and Immersive VR Audio Experience Using Scene Based Audio/Higher Order Ambisonics." *Proceedings of the Audio Engineering Society International Conference on Audio for Virtual and Augmented Reality* (Los Angeles, CA).

Simon, Germain, Sean E. Olive, and Todd Welti. 2009. "The Effect of Whole-Body Vibrations on Preferred Bass Equalization of Automotive Audio Systems." *127th Convention of the Audio Engineering Society* Convention Paper 7956.

Slater, Mel, and Sylvia Wilbur. n.d. "A Framework for Immersive Virtual Environments (FIVE): Speculations on the Role of Presence in Virtual Environments." *Presence* 6 (6): 603–16.

Slater, Mel, Domna Banakou, Alejandro Beacco, Jaime Gallego, Francisco Macia-Varela, and Ramon Oliva. 2022. "A Separate Reality: An Update on Place Illusion and Plausibility in Virtual Reality." *Frontiers in Virtual Reality* 3.

Slater, Mel. 2009. "Place Illusion and Plausibility Can Lead to Realistic Behaviour in Immersive Virtual Environments." *Philosophical Transactions of the Royal Society B: Biological Sciences* 364 (1535): 3549–57.

Sound on Sound. 2022. "Dolby Announce Personalised HRTF App," March 17, 2022. https://www.soundonsound.com/news/dolby-announce-personalised-hrtf-app.

Speicher, Maximilian, Brian D Hall, and Michael Nebeling. 2019. "What Is Mixed Reality?" In *CHI '19: Proceedings of the 2019 CHI Conference on Human Factors in Computing Systems*, 537:1–15.

Steinberg, J. C., and W. B. Snow. 1934. "Auditory Perspective—Physical Factors." *Electrical Engineering*, 12–17.

Su, Hengwei, Atsushi Marui, and Toru Kamekawa. 2019. "The Effect of HRTF Individualization and Head-Tracking on Localization and Source Width Perception in VR." In *Proceedings of the 146th Audio Engineering Society (AES) Convention*. Dublin, Ireland.

Sunder, Kaushik, Saarish Kareer, Scott Murakami, and Axel Borgmo. 2023. "Personalized Spatial Audio Tools for Immersive Audio Production and Rendering." *2023 Immersive and 3D Audio: From Architecture to Automotive (I3DA)*, September 5, 1–10. https://doi.org/10.1109/I3DA57090.2023.10289506.

Sutherland, Ivan E. 1965. "The Ultimate Display." In *Proceedings of the IFIP Congress*, 2:506–8. New York.

Tak, Willem. 1958. "The 'Electronic Poem' Performed in the Philips Pavilion at the 1958 Brussels World Fair: B. The Sound Effects." *Philips Technical Review* 20 (2/3): 43–44.

Tarkovsky, Andrey. 1989. *Sculpting in Time: Reflections on the Cinema*. Translated by Kitty Hunter-Blair. University of Texas Press.

Taruffi, Liila, and Mats B. Küssner. 2019. "A Review of Music-Evoked Visual Mental Imagery: Conceptual Issues, Relation to Emotion, and Functional Outcome." *Psychomusicology: Music, Mind, and Brain* 29 (2–3): 62.

Theile, Gunther, and Georg Plenge. 1977. "Localization of Lateral Phantom Sources." *Journal of the Audio Engineering Society* 25 (4): 196–200.

Theile, Gunther. 2001. "Natural 5.1 Music Recording Based on Psychoacoustic Principles." Paper presented at 19th Audio Engineering Society International Conference: Surround Sound - Techniques, Technology, and Perception, Elmau, Germany. *19th Audio Engineering Society International Conference: Surround Sound – Techniques, Technology, and Perception.*

Todd, Craig C., Grant A. Davidson, Mark F. Davis, Louis D. Fielder, Brian D. Link, and Steve Vernon. 1994. "AC-3: Flexible Perceptual Coding for Audio Transmission and Storage." Amsterdam, The Netherlands: Presented at the *96th Convention of the Audio Engineering Society*, Convention Paper 3796.

Torick, Emil. 1998. "Highlights in the History of Multichannel Sound." *Journal of the Audio Engineering Society* 46 (1/2): 27–31.

Turchet, Luca, Rob Hamilton, and Anıl Çamcı. 2021. "Music in Extended Realities." *IEEE Access* 9: 15810–32.

Vaggione, Horacio. 2001. "Composing Musical Spaces by Means of Decorrelation of Audio Signals." In *Proceedings of the COST G-6 Conference on Digital Audio Effects (DAFx-01)*, Addendum 1–8. Limerick, Ireland.

Ventura, Sara, Laura Badenes-Ribera, Rocio Herrero, Ausias Cebolla, Laura Galiana, and Rosa Baños. 2020. "Virtual Reality as a Medium to Elicit Empathy: A Meta-Analysis." *Cyberpsychology, Behavior, and Social Networking* 23 (10): 667–76.

Vilaplana Stark, Matias, and Anıl Çamcı. 2024. "Unforeseen Collisions: A Networked VR System for Musical Collaboration." In *Proceedings of the 5th AES International Conference on Audio for Virtual and Augmented Reality*. Redmond, Washington: Audio Engineering Society.

Vorländer, Michael. 2020a. *Auralization: Fundamentals of Acoustics, Modelling, Simulation, Algorithms and Acoustic Virtual Reality*. 2nd ed. Springer Cham. https://doi.org/10.1007/978-3-030-51202-6.

Vorländer, Michael. 2020b. *Auralization: Fundamentals of Acoustics, Modelling, Simulation, Algorithms and Acoustic Virtual Reality*. 2nd ed. Springer Cham. https://doi.org/10.1007/978-3-030-51202-6.

Wallach, Hans, Edwin B. Newman, and Mark R. Rosenzweig. 1949. "The Precedence Effect in Sound Localization." *American Journal of Psychology* 62 (3): 315–36.

Warnecke, Michaela, Samuel Clapp, Ben-Hur Zamir, David Lou Alon, Sebastià V. Amengual Garí, and Paul Calamia. 2024. "Sound Sphere 2: A High-Resolution HRTF Database." In *AES 5th International Conference on Audio for Virtual and Augmented Reality, Conference Paper 31*. Redmond, Washington: Audio Engineering Society.

Warren Jr, William H., Elizabeth E Kim, and Robin Husney. 1987. "The Way the Ball Bounces: Visual and Auditory Perception of Elasticity and Control of the Bounce Pass." *Perception* 16 (3): 309–36.

Warusfel, Olivier. 2003. "Listen HRTF Database." IRCAM. http://recherche.ircam.fr/equipes/salles/listen/.

Wenzel, Elizabeth M., Durand R. Begault, and Martine Godfroy-Cooper. 2018. "Perception of Spatial Sound." In *Immersive Sound: The Art and Science of Binaural and Multi-Channel Audio*, edited by Agnieszka Roginska and Paul Geluso. Routledge.

Wightman, Frederic L., and Doris J. Kistler. 1989. "Headphone Simulation of Free-Field Listening. II: Psychophysical Validation." *Journal of the Acoustical Society of America* 85 (2): 868–78.

Willette, Aaron, Nachiketa Gargi, Eugene Kim, Julia Xu, Tanya Lai, and Anıl Çamcı. 2020. "Cross-Platform and Cross-Reality Design of Immersive Sonic Environments." In *Proceedings of the International Conference on New Interfaces for Musical Expression*, 127–30.

Wittek, Helmut, and Gunther Theile. 2017. "Development and Application of a Stereophonic Multichannel Recording Technique for 3D Audio and VR." *143rd Convention of the Audio Engineering Society*, Paper Number: 9869.

Woszczyk, Wieslaw, and Paul Geluso. 2018. "Streamlined 3D Sound Design: The Capture and Composition of a Sound Field." *145th Convention of the Audio Engineering Society*, Convention e-Brief 459.

Woszczyk, Wieslaw, Søren Bech, and Villy Hansen. 1995. "Interactions Between Audio-Visual Factors in a Home Theater System: Definition of Subjective Attributes." *99th Convention of the Audio Engineering Society* Preprint 4133.

Yang, Jing, Amit Barde, and Mark Billinghurst. 2022. "Audio Augmented Reality: A Systematic Review of Technologies, Applications, and Future Research Directions." *Journal of the Audio Engineering Society* 70 (10): 788–809.

Yeo, Sang Seok, Jung Won Kwon, and Seo Yoon Park. 2022. "EEG-Based Analysis of Various Sensory Stimulation Effects to Reduce Visually Induced Motion Sickness in Virtual Reality." *Scientific Reports* 12 (1): 18043.

Index

For Product Safety Concerns and Information please contact our EU
representative GPSR@taylorandfrancis.com
Taylor & Francis Verlag GmbH, Kaufingerstraße 24, 80331 München, Germany

www.ingramcontent.com/pod-product-compliance
Ingram Content Group UK Ltd.
Pitfield, Milton Keynes, MK11 3LW, UK
UKHW050229200926
475702UK00020B/2929
* 9 7 8 0 3 6 7 4 9 8 9 7 9 *